NEW

Co-ordinated
SCIENCE

Physics

second edition

Stephen Pople   Peter Whitehead

Oxford University Press

Oxford University Press, Great Clarendon Street, Oxford OX2 6DP

Oxford New York
Athens Auckland Bangkok Bogotá Buenos Aires
Calcutta Cape Town Chennai Dar es Salaam
Delhi Florence Hong Kong Istanbul Karachi
Kuala Lumur Madrid Melbourne Mexico City
Mumbai Nairobi Paris São Paulo Singapore
Taipei Tokyo Toronto Warsaw

and associated companies in
Berlin Ibadan

Oxford is a trade mark of Oxford University Press

© Stephen Pople 1996
Peter Whitehead 1996

First Published 1989 (ISBN 0 19 914247 5)
Reprinted 1990, 1991 (twice), 1992, 1993 (twice), 1994

2nd edition published 1996. Reprinted 1996, 1997 (twice), 1998 , 1999
ISBN 0 19 914651 9   School edition
ISBN 0 19 914672 1   Bookshop edition

Typeset in 10½/13pt Photina by Acc Computing Limited.
Queen Camel, Somerset
Printed in Spain

Additional photographs by Peter Gould and Chris Honeywell. With
special thanks to F C Bennett & Sons Ltd, Smiths Security Services,
The Straw Hat Bakery and N J Thake Cycles.

Illustrations by Nick Hawken and Associates, Clive Goodyer and Jan
Lewis.

Introduction

This book deals with physics, its practical uses and some of the social issues it raises. You are most likely to find the book useful if you are studying physics as part of a GCSE Co-ordinated Science course or as a single GCSE subject.

To help you find things more easily, the book is written in two-page units, each dealing with a different topic. The units are grouped into sections.
- **Use the contents page** if you want to see the main headings.
- **Use the index** at the back if you want to look up information about one particular thing. The index is alphabetical.
- **Use the questions** to test yourself. There are questions covering the main ideas at the end of each unit, and there are examination-level questions at the end of each section.
- **Use the reference section** near the back of the book if you need to look up the following:

   Units of measurement
   Electrical symbols
   Answers to numerical questions
   The scale of the Universe

Physics is an important subject. It doesn't just happen in laboratories. It is all around you, in fairgrounds, fields, farms, and factories. It is taking place deep in the Earth and far out in space. You'll find physics everywhere.

Stephen Pople
Peter Whitehead   January 1996

Contents

1.1 Units for measuring

SI units – a common system of measuring

Which units would you use to measure:

- a length?
- or a mass?
- or a time?

There are several possibilities. But in scientific work, life is much easier if everyone uses the same system of units. Nowadays, most scientists use the **SI system** (from the French, Système International d'Unites). This starts with the metre, the kilogram and the second. Many other units are based on these.

Length

The **metre** (**m**, for short) is the SI unit of length.

The chart shows some of the larger and smaller length units based on the metre.

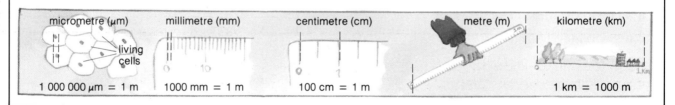

micrometre (μm)	millimetre (mm)	centimetre (cm)	metre (m)	kilometre (km)
living cells				
1 000 000 μm = 1 m	1000 mm = 1 m	100 cm = 1 m		1 km = 1000 m

Mass

Mass is the amount of matter in something. In the laboratory, it is often measured using a top pan balance. Mass is sometimes called 'weight'. This is wrong. The difference between mass and weight is explained on page 24.

The **kilogram** (**kg**) is the SI unit of mass.

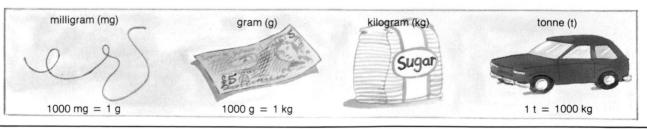

| milligram (mg) | gram (g) | kilogram (kg) | tonne (t) |
| 1000 mg = 1 g | 1000 g = 1 kg | | 1 t = 1000 kg |

Time

The **second** (**s**) is the SI unit of time.

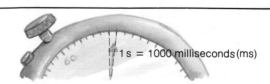

1 s = 1000 milliseconds (ms)

Volume

Volume is the amount of space something takes up. The SI unit of volume is the **cubic metre (m³)**. This is rather a large unit of volume for everyday use, so the litre, millilitre, or centimetre cubes are often used instead.

There are 1000 millilitres in 1 litre, and 1000 cubic centimetres in 1 litre. So 1 millilitre is the same as 1 cubic centimetre.

Measuring volume

If something has a simple shape, its volume can be calculated. For example:

volume of a
rectangular block = length × width × height

Liquid volumes of about a litre or less can be measured using a measuring cylinder. Pour in the liquid, and the reading on the scale gives the volume.

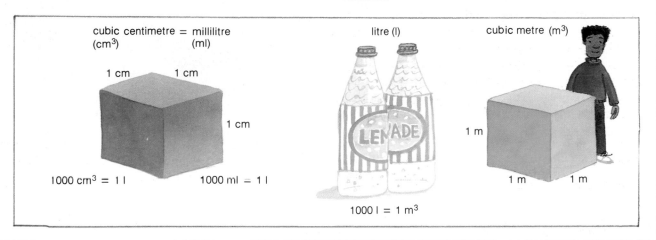

cubic centimetre = millilitre (cm³) (ml)

1 cm   1 cm

1 cm

1000 cm³ = 1 l   1000 ml = 1 l

litre (l)

LEMADE

cubic metre (m³)

1 m

1 m   1 m

1000 l = 1 m³

---

1 Copy and complete the table:

	Unit	Abbreviation
Length	?	m
?	kilogram	?
Time	?	?

2 What do the following stand for:
mm, t, mg, ms, l, cm?

3 Which is the greater?
1600 g or 1.5 kg?
1450 mm or 1.3 m?

4 10, 100, 1000, 100 000, 1 000 000:
Which of these is:
a the number of mg in 1 g?
b the number of mm in 1 cm?
c the number of cm in 1 km?
d the number of cm in 1 m?
e the number of mm in 1 km?

5 Write down the value of:
a 1 m in mm
b 1.5 m in mm
c 1.534 m in mm
d 1652 mm in m

6 Write down the values of:
a 2.750 m in mm   **b** 1.600 km in m
c 6.500 g in mg   **d** 150 cm in m
e 1750 g in kg

7 Which is the odd one out in each of the boxes?

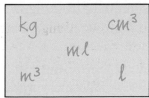

kg   cm³
    ml
m³   l

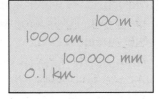

                100 m
1000 cm
                100 000 mm
0.1 km

8 What is the volume of the liquid in the measuring cylinder in the diagram?
What is the volume
– in cm³?
– in litres?

9 What is the volume of a metal block 3 cm long by 2 cm wide by 4 cm high?
What would be the volume of a block twice as long, wide and high?

ml
1000

500

1.2 Density

Is lead heavier than water? Not necessarily. It depends how much lead and water you are comparing. However, lead is more dense than water – it has more kilograms packed into every cubic metre.

You can calculate the density of a material if you know its mass and its volume:

$$\text{density} = \frac{\text{mass}}{\text{volume}}$$

Take the case of water:
a mass of 1000 kg has a volume of 1 m³;
a mass of 2000 kg has a volume of 2 m³;
a mass of 3000 kg has a volume of 3 m³;
and so on.

Using any of these sets of figures in the above equation:
the density of water works out to be 1000 kg/m³. There are 1000 kilograms of water packed into every cubic metre.

The kg/m³ is the basic SI unit of density. But it isn't always the easiest unit to use in laboratory work. When masses are measured in grams and volumes in centimetre cubed, it is simpler to calculate densities in g/cm³. Changing to kg/m³ is easy: 1 g/cm³ = 1000 kg/m³.

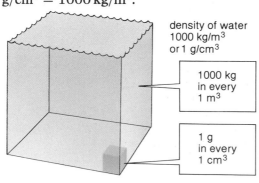

density of water 1000 kg/m³ or 1 g/cm³

1000 kg in every 1 m³

1 g in every 1 cm³

How dense?

The least dense material you are ever likely to see from Earth is the dimly glowing gas in the tail of a comet. The gas stretches for millions of miles behind the comet's head. It is so thin, that there is less than a kilogram of gas in each kilometre cubed.

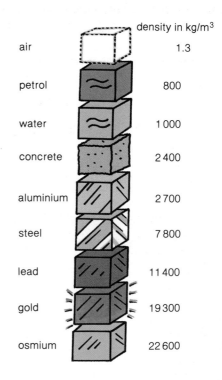

	density in kg/m³
air	1.3
petrol	800
water	1 000
concrete	2 400
aluminium	2 700
steel	7 800
lead	11 400
gold	19 300
osmium	22 600

The rare metal osmium is the densest substance found on Earth. It is about twice as dense as lead. If this book were made of osmium, it would weigh as much as a television set.

Density calculations

Start with the equation:

$$\text{density} = \frac{\text{mass}}{\text{volume}}$$

Then rearrange it:

$$\text{mass} = \text{volume} \times \text{density}$$

and:

$$\text{volume} = \frac{\text{mass}}{\text{density}}$$

These are useful if you know the density of something, but want to find its mass or volume. This triangle may help you to remember all three equations – cover 'Volume' if you want the equation for volume, and so on:

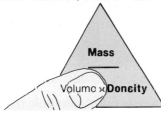

Example *A block of aluminium has a volume of 2 m³ and a density of 2700 kg/m³. What is its mass?*

Using the triangle: mass = volume × density

so, $\text{mass} = 2\,\text{m}^3 \times 2700\,\text{kg/m}^3$

$= 5400\,\text{kg}$

The mass of the aluminium is 5400 kg.

Brewing with gravity

Ask brewers how strong their beer is, and they might tell you that it has a 'gravity' of 1020°. Their scale of gravity is actually a scale of density, because beers become more dense as they get stronger. Plain water is the weakest 'beer' of all. This has a gravity of 1000° on the brewer's scale.

	Gravity	Density
WATER	1000°	1000 kg/m³
WEAK BEER	1010°	1010 kg/m³
STRONG BEER	1040°	1040 kg/m³

Assume $g = 10\,\text{N/kg}$.
To answer these questions, you will need to use some of the density values in the chart on the opposite page.

1 Work out:
a the mass of water needed to fill a 4 m³ tank;
b the volume of a storage tank which will hold 1600 kg of petrol;
c the mass of air in a room measuring 5 m × 2 m × 3 m.

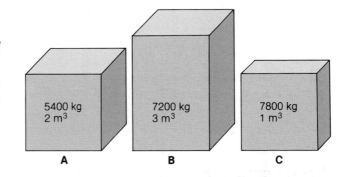

2 Which block has:
a the greatest mass; **b** the greatest volume;
c the greatest density?
Use the chart to decide which material each block could be made from.

3 A builder wants to load a stack of 100 solid concrete blocks on his lorry.

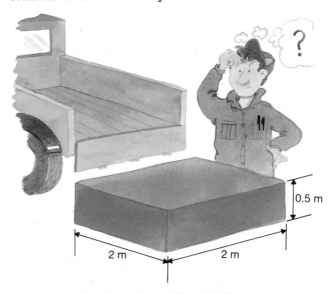

The stack measures 2 m × 2 m × 0.5 m. Work out:
a the volume of the concrete;
b the mass of the concrete;
c whether he can safely carry all the blocks on the lorry, if the maximum load is 3000 kg. And if not, how many blocks he can carry.

1.3 Measuring density

Who needs to?

Airline pilots need to know air density before they take off.

Geologists use density measurements to help identify types of rock.
This information is useful to . . .

Brewers, like milk inspectors use density checks to help maintain quality.

. . . **Civil engineers** who must calculate the mass of their building materials, and decide if the ground can support the weight.

250 BC: A Greek Scientist who Discovered how to Investigate Density

The King of Syracuse gave his goldsmith some gold. 'Make a crown', he said. So the goldsmith did. But the king was suspicious. Perhaps the goldsmith had kept some of the gold. Perhaps he had mixed in cheaper silver to take its place. He asked Archimedes to test the crown.

Archimedes knew that the crown was the correct mass. He knew that silver was less dense than gold. So, if the crown contained any silver, it would have more volume than it should. But how could he check the volume of the crown? Stepping into his bath one day, Archimedes noticed the rise in water level. Here was the answer to the problem! Put the crown in the water and measure the rise in level. Put an equal mass of pure gold in the water. See if the rise in level was the same. 'Eureka', he shouted. Which is Greek for 'I have found it'.

Was the crown made of pure gold? Unfortunately, nobody wrote down the result. So we shall never know . . .

How to measure a density

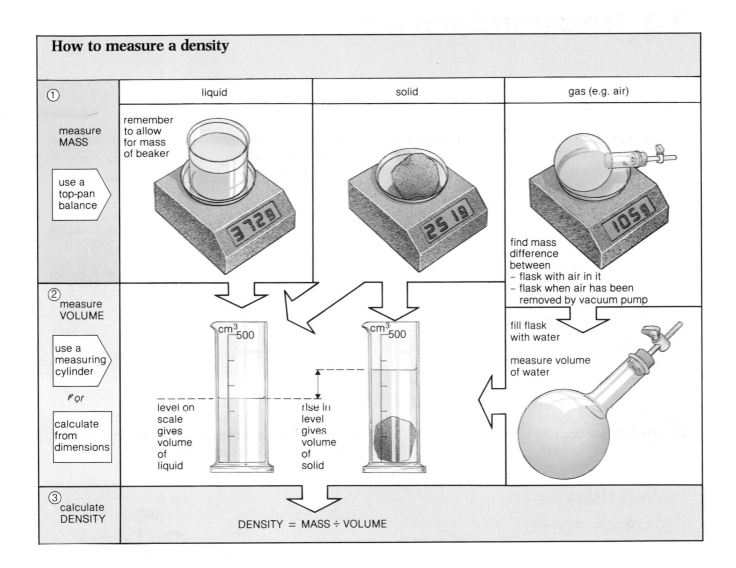

	liquid	solid	gas (e.g. air)
① measure MASS — use a top-pan balance	remember to allow for mass of beaker		find mass difference between – flask with air in it – flask when air has been removed by vacuum pump
② measure VOLUME — use a measuring cylinder ♩ or calculate from dimensions	level on scale gives volume of liquid	rise in level gives volume of solid	fill flask with water — measure volume of water
③ calculate DENSITY	DENSITY = MASS ÷ VOLUME		

1

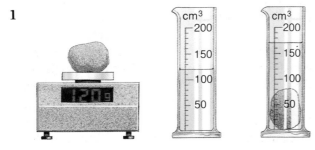

What is:
a the volume of the stone in cm³?
b the density of the stone in g/cm³?

2

Volume of liquid in beaker	= 200 cm³
Mass of beaker	= 110 g
Mass of beaker, filled with liquid	= 270 g

What is:
a the mass of the liquid?
b the density of the liquid in g/cm³?

3 Solve the same problem as Archimedes using the information given here – no need to take a bath!

	A	B	C
mass in g	3800	3800	3800
Volume in cm³	380	200	300

The density of gold is 19 g/cm³.

The density of silver is 10 g/cm³.

Decide which crown is gold, which is silver, and which is a mixture.

1.4 On the move

Speed

The police can check the speed of a car with a radar 'gun'. But there's a simpler method. Measure the distance between two points along a road – say, two lamp posts. Measure the time a car takes to travel between these points. Then calculate the speed:

$$\text{average speed} = \frac{\text{distance moved}}{\text{time taken}}$$

For example, a car which travels 50 metres in 5 seconds, has an average speed of 10 metres per second – written 10 m/s for short.

On most journeys, the speed of a car varies, so the actual speed at any moment is usually different from the average speed. To find an actual speed, you have to find the distance travelled in the shortest time you could measure.

Velocity

As with speed, the unit of velocity is the metre per second (m/s). Velocity tells you the speed at which an object is travelling. But it also tells you the direction of travel:

$$\text{average velocity} = \frac{\substack{\text{distance moved} \\ \text{in a particular direction}}}{\text{time taken}}$$

In diagrams you can show a velocity with an arrow:

$$\xrightarrow{\hspace{1cm} 10\,\text{m/s} \hspace{1cm}}$$

Or you can use a + or − can to give the direction. For example:

$$+10\,\text{m/s} \quad (10\,\text{m/s to the right})$$
$$-10\,\text{m/s} \quad (10\,\text{m/s to the left})$$

Quantities like velocity which have a direction as well as a value are called **vectors**.

What they mean		
A steady speed of 10 m/s	A distance of 10 metres is travelled every second	
A steady velocity of + 10 m/s	A distance of 10 metres is travelled every second (to the right)	
A steady acceleration of 5 m/s²	Speed goes up by 5 metres/second every second	
A steady retardation of 5 m/s²	Speed goes down by 5 metres/second every second	

Acceleration

From a standing start, a rally car can reach a velocity of 50 m/s in 10 s or less. It gains velocity very rapidly. It has a high **acceleration**.

Like velocity, acceleration is a vector. It is calculated as follows:

$$\text{acceleration} = \frac{\text{gain in velocity}}{\text{time taken}}$$

For example, if a car gains an extra 50 m/s of velocity in 10 seconds:

$$\text{acceleration} = \frac{50}{10} \text{ m/s per second}$$

$$= 5 \text{ m/s per second}$$

which is written as 5 m/s² for short.

Retardation is the opposite of acceleration. If a car has a retardation of 5 m/s², it is *losing* 5 m/s of speed every second.

1 A car travels 500 m in 20 s, what is its average speed?
Why is the actual speed of a car not usually the same as its average speed?
2 How far does
the car travel in
1 s? 5 s? 10 s?

10 m/s

How long does it take to travel 90 m?
3 Copy and complete:
A motor cycle has a steady _____ of 3 m/s². This means that every __ its _____ increases by _____.
4 A car has steady acceleration. The chart shows how its speed increases. Copy and complete the chart.

After	1 s	2 s	3 s	4 s	5 s	?
Speed	4 m/s	8 m/s	?	16 m/s	?	28 m/s
Steady acceleration = ? m/s²						

5 An aircraft on its take-off run has a steady acceleration of 3 m/s².
How much velocity does it gain in 10 s?
If the aircraft has a velocity of 20 m/s as it passes a post, what is its velocity 10 s later?
6 A motor cycle takes 8 s to increase its velocity from 10 m/s to 30 m/s. What is its average acceleration?
7 A rally driver has 5 s to stop her car, which is travelling at a speed of 20 m/s. What is her average retardation?

How to do calculations

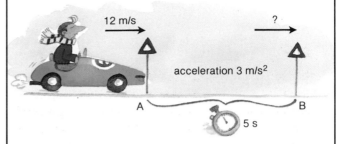
12 m/s   ?
acceleration 3 m/s²
A   B
5 s

The car in the diagram passes point A with a velocity of 12 m/s.
The car has an acceleration of 3 m/s².
What is the velocity of the car as it passes point B 5 s later?

The car is gaining an extra 3 m/s of velocity every second.

So, in 5 s, it gains an extra 15 m/s of velocity on top of its original velocity of 12 m/s.

Its final speed on passing B is therefore 12 m/s + 15 m/s, which equals 27 m/s.

This could be written in another way:
Final velocity = original velocity + extra velocity
or
Final velocity
= original velocity + (acceleration × time)

It sometimes helps to remember the above equation in symbols:

$v = u + at$
where
v is the final velocity, a is the acceleration,
u is the original velocity,
t is the time taken.

In the case of the car,

$$v = 12 + (5 \times 3) \quad \text{m/s}$$
$$= 12 + 15 \qquad \text{m/s}$$
$$= 27 \text{ m/s}$$

The equation works for retardation as well. Just call the retardation a negative acceleration. For example, a retardation of 5 m/s² is an acceleration of -5 m/s².

1.5 Motion graphs

You can learn a lot from motion graphs. They can tell you how far something has travelled, how fast it is moving, and all the speed changes there have been.

Distance–time graphs

Imagine a car travelling along a road. There is a post on the road. Every second, the distance of the car from the post is measured. Distance and time readings are recorded in a chart, and used to plot a graph.

Here are the results from just four possible journeys. One is hardly a journey at all.

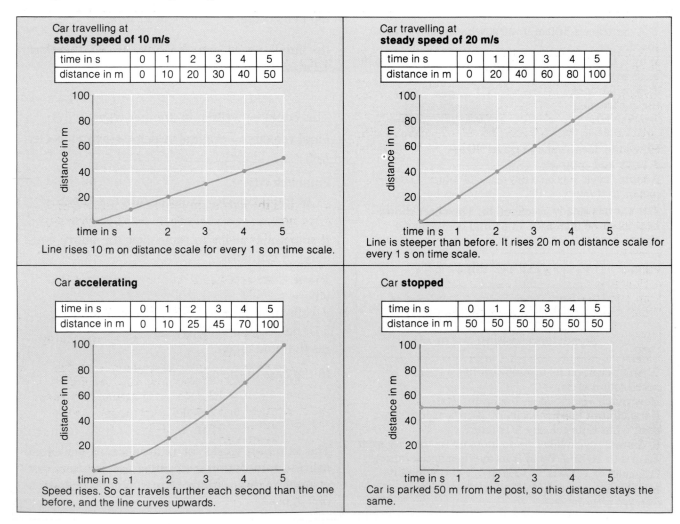

Car travelling at **steady speed of 10 m/s**

time in s	0	1	2	3	4	5
distance in m	0	10	20	30	40	50

Line rises 10 m on distance scale for every 1 s on time scale.

Car travelling at **steady speed of 20 m/s**

time in s	0	1	2	3	4	5
distance in m	0	20	40	60	80	100

Line is steeper than before. It rises 20 m on distance scale for every 1 s on time scale.

Car **accelerating**

time in s	0	1	2	3	4	5
distance in m	0	10	25	45	70	100

Speed rises. So car travels further each second than the one before, and the line curves upwards.

Car **stopped**

time in s	0	1	2	3	4	5
distance in m	50	50	50	50	50	50

Car is parked 50 m from the post, so this distance stays the same.

Speed–time graphs

Don't confuse these with distance–time graphs. The shapes may look the same, but their meaning is very different.

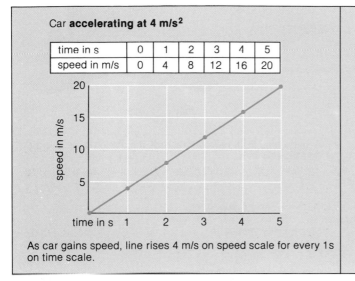

Car accelerating at 4 m/s²

time in s	0	1	2	3	4	5
speed in m/s	0	4	8	12	16	20

As car gains speed, line rises 4 m/s on speed scale for every 1 s on time scale.

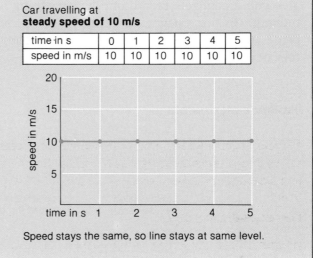

Car travelling at steady speed of 10 m/s

time in s	0	1	2	3	4	5
speed in m/s	10	10	10	10	10	10

Speed stays the same, so line stays at same level.

1 A motor cycle passes a lamp post. Every second, its distance from the post is measured:

time/s	0	1	2	3	4	5	6	7	8	9
distance/m	0	3	10	22	34	46	54	56	56	56

a Plot a distance–time graph
b Mark on your graph the sections where the motor cycle:
has acceleration; is travelling at a steady speed; has retardation; is stopped.
c How far does the motor cycle travel in the first 7 seconds?
What is its average speed over this period?
d How long does it take the motor cycle to travel from 10 m to 46 m? What distance does it cover? What is its average speed over this distance?

2 The graph shows a speed–time graph for another motor cycle.

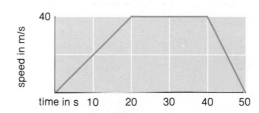

a What is the maximum speed of the motor cycle?
b For how many seconds does the motor cycle stay at its maximum speed?
c For how many seconds is the motor cycle actually moving?
d How much speed does the motor cycle gain in the first 20 seconds?
How much speed does it gain every second?
What is its acceleration?
e What is the retardation of the motor cycle during the last 10 seconds?
3 It takes a driver 10 minutes to get to work. She stops to buy a paper on the way, has a set of traffic lights on her route, and a short section of motorway. Sketch a typical speed–time graph for her journey, and don't let her break any speed limits.
(1 m/s is about 2 mph.)
4 An ice skater skates once round an oval rink. He accelerates down the length of the rink, but takes the ends more slowly. He falls over once. Sketch a speed-time graph for him.

1.6 Timing with tape

A piece of paper tape is rather like a graph. It can give you a complete record of how something is moving. On these pages, however, it isn't a car being studied, but a trolley moving on a laboratory bench.

A 'black box' flight recorder records the motion of an aircraft on magnetic wire.

Trolley experiments

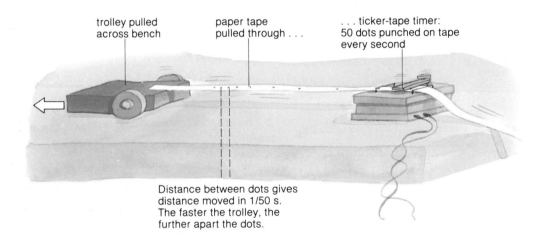

trolley pulled across bench

paper tape pulled through . . .

. . . ticker-tape timer: 50 dots punched on tape every second

Distance between dots gives distance moved in 1/50 s. The faster the trolley, the further apart the dots.

Examples of tapes

start

steady speed: distance between dots stays the same

higher steady speed: distance between dots greater than before

acceleration: distance between dots increases

acceleration — — — — — — — then — — — — — — — retardation

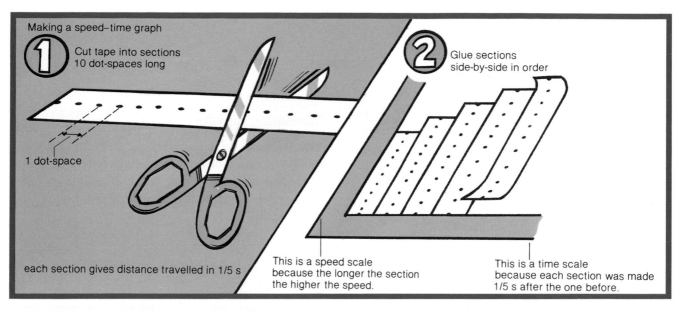

Making a speed–time graph

1 Cut tape into sections 10 dot-spaces long

1 dot-space

each section gives distance travelled in 1/5 s

2 Glue sections side-by-side in order

This is a speed scale because the longer the section the higher the speed.

This is a time scale because each section was made 1/5 s after the one before.

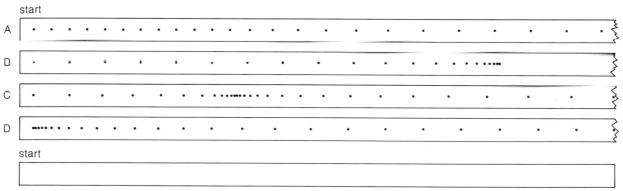

start

A

D

C

D

start

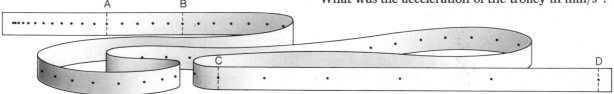

1 Which of the tapes above shows:
a acceleration, then a steady speed?
b retardation until stopped, then acceleration?
c a steady speed, then acceleration, then a higher steady speed?

2 A trolley travelling at a steady speed, loses speed, stops, then accelerates. Copy the blank tape above. Mark in the pattern of dots you might expect to see.

3 In the questions which follow, you have to make measurements on the paper tape shown below. The ticker-tape timer made 50 dots on paper tape every second. The distance from one dot to the next is called a *dot-space*.

a How long did it take the timer to make 5 dot-spaces?
b How many dot-spaces are there between A and B?
c How long did it take the tape to move from A to B?
d Use a mm ruler to measure the distance from A to B.
e What was the average speed (in mm/s) of the trolley between A and B?
f Measure the distance from C to D, then work out the average speed of the trolley between C and D.
g Section CD was completed exactly one second after section AB.
By how much did the speed of the trolley increase in this time?
h An acceleration of 1 mm/s² means that the speed increases by 1 mm/s every second.
What was the acceleration of the trolley in mm/s²?

1.7 Falling freely

When Flight Lieutenant Alkemade first left his aircraft, he fell towards the ground with an acceleration of 10 m/s². Rushing air quickly reduced this acceleration. Otherwise, he would have hit the ground at over 700 mph (about 300 m/s).

Air resistance slows down some things more than others. It doesn't slow a falling rock very much. But it slows a feather a lot. Without air resistance, all things falling near the Earth would have the same acceleration of 10 m/s². This is the **acceleration of free fall**, g.

So, without air resistance, anything dropped would speed up like this:

0 s   no speed

after 1 s   10 m/s

after 2 s   20 m/s

after 3 s   30 m/s

acceleration of free fall g is 10 m/s² for all objects

and so on

Calculating how far

With the value of g, and the time something takes to fall, you can work out the height fallen. Here is a problem about a falling stone. The problem is done in two ways – with actual values, and with letter symbols.

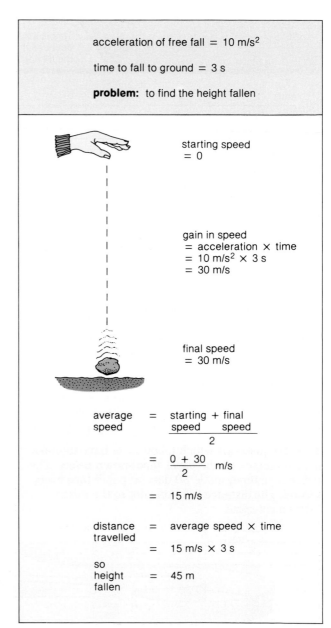

acceleration of free fall = 10 m/s²

time to fall to ground = 3 s

problem: to find the height fallen

starting speed = 0

gain in speed
= acceleration × time
= 10 m/s² × 3 s
= 30 m/s

final speed = 30 m/s

$$\text{average speed} = \frac{\text{starting speed} + \text{final speed}}{2}$$

$$= \frac{0 + 30}{2} \text{ m/s}$$

$$= 15 \text{ m/s}$$

distance travelled = average speed × time

= 15 m/s × 3 s

so height fallen = 45 m

Measuring *g*

First rearrange the equation for *h* below to give an equation for *g*:

$$g = \frac{2h}{t^2}$$

Drop a metal ball. Measure the height (*h*) it falls in metres. Measure the time (*t*) it takes to fall in seconds. Then use the new equation to work out *g*.

Careful measurements show that *g* has a value of 9.8 m/s². However, the value of 10 m/s² is simpler to use, and accurate enough for most calculations.

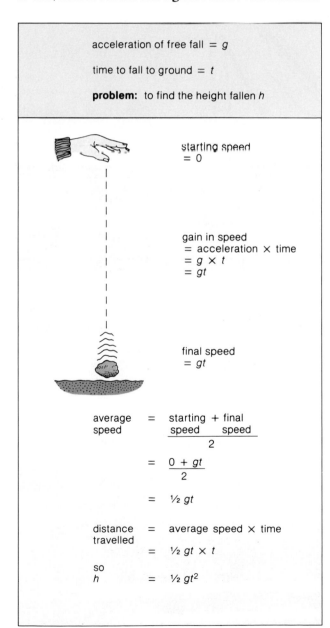

acceleration of free fall = *g*

time to fall to ground = *t*

problem: to find the height fallen *h*

starting speed = 0

gain in speed
= acceleration × time
= *g* × *t*
= *gt*

final speed = *gt*

$$\text{average speed} = \frac{\text{starting speed} + \text{final speed}}{2}$$

$$= \frac{0 + gt}{2}$$

$$= \tfrac{1}{2}gt$$

distance travelled = average speed × time

$$= \tfrac{1}{2}gt \times t$$

so

$$h = \tfrac{1}{2}gt^2$$

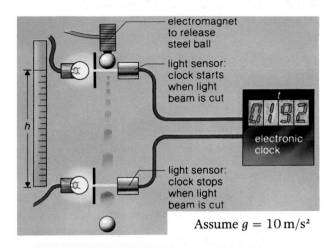

electromagnet to release steel ball

light sensor: clock starts when light beam is cut

t

electronic clock

light sensor: clock stops when light beam is cut

Assume *g* = 10 m/s²

1 Is each of the following TRUE or FALSE?
Without air resistance:
a A heavy stone falls more quickly than a light stone.
b Near the ground, falling things all accelerate at the same rate.
c Dropped from the same height, a heavy stone takes exactly the same time to reach the ground as a light stone.

2 Copy the chart, filling in the missing information about the falling stone:

time in s	0	1s	2s		?	?
speed in m/s	0	10	?		40	50
speed gained every second = ?						
acceleration = ?						

3 The metal ball below has been falling for 5 seconds.
What is its acceleration?
What is its speed?
What is its average speed?
What height has it fallen?
How far will the ball have fallen in the first 10 seconds?

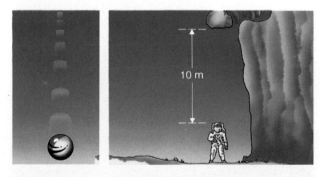

10 m

4 The astronaut is on a planet where the acceleration of free fall is only 1 m/s². It takes him 4 seconds to move sideways. Will he be able to escape the falling rock?

1.8 Thrills and spills

These cars can reach 60 mph in just three seconds. That's an acceleration of 10 m/s², the same as the acceleration of free fall, g.
If you were in the driving seat, you would feel the acceleration as a push in the back – a push as strong as your own weight.

These rides give you the effects of high acceleration by making you travel round tight bends very fast.

This roller-coaster ride lasts just over a minute. You experience speeds up to 25 m/s (50 mph), and accelerations up to 3 g.

High acceleration can drain blood from your head to your feet and make you 'black out'. But not during this ride. Sitting with your knees up stops the rush of blood to your feet.

A 3 g ride. You may feel most scared at the high point of the swing. But it's near the low point that you are most firmly pushed into your seat.

For this 5 g ride, you need skill and a million pounds worth of training. And a special suit which squeezes your limbs tightly to reduce the flow of blood from your head.

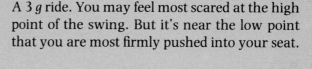

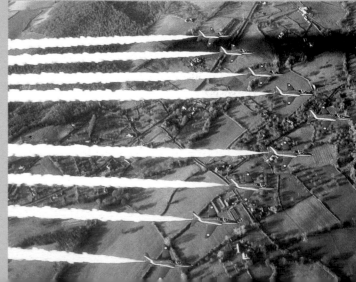

If another car crashes into the back of yours, a headrest can save you from serious neck injury. It makes sure that your head accelerates forward at the same rate as the rest of your body.

Why not build them stronger?
In a crash, it's safer for the passengers if the front of the car *does* collapse. It means lower deceleration and less risk of injury. But the metal body must form a strong 'cage' round the passengers to stop them being crushed.

Most thrill machine rides have plenty of built-in safety. If one metal part breaks, there are still others left to support you. But how many fairground rides can you think of where your safety depends on the strength of just one metal part?

If rear-facing seats are safer, why has no airline decided to 'go it alone' and fit them? Can you suggest reasons?

In thrill machine rides, you can experience accelerations of 3 g or more. But without headrests, accelerations like this wouldn't be safe. Why not?

The safest way to travel – backwards.
If a plane makes a crash landing, the deceleration can be very high. Rear-facing seats give the best chance of survival. Which is why the RAF fit them to their transport aircraft. But airlines haven't taken up the idea.

1.9 Force
– the secret of
acceleration

A force is a push or a pull. It is a vector because it has a direction.

Rolls Royce engines use a test rig to measure one rather large force – the forward thrust from a jet engine.

Like all forces, the force from a jet engine is measured in a unit called the **newton** (**N**). Some typical force values are:

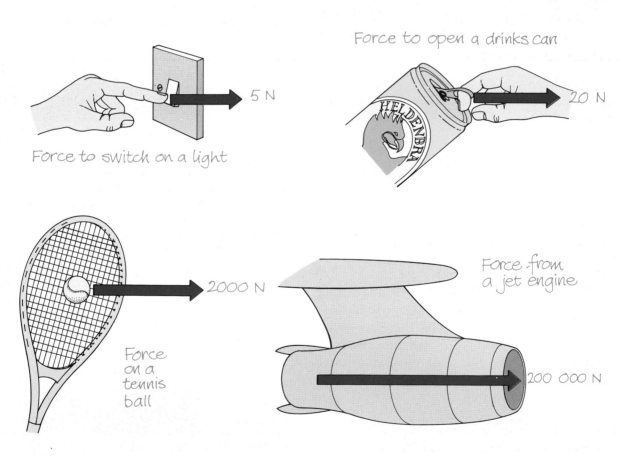

Force to switch on a light — 5 N

Force to open a drinks can — 20 N

Force on a tennis ball — 2000 N

Force from a jet engine — 200 000 N

Small forces can be measured with a **spring balance**. The greater the force, the more the spring is then stretched, and the higher the reading on the scale.

Resisting acceleration

It takes over an hour for an ocean-going tanker to reach full speed – and over an hour for it to be stopped when the engines are put in reverse. Like all masses, the tanker resists any change in velocity. The effect is called **inertia**. The more mass something has, the greater its inertia, and the more it resists acceleration.

Making mass accelerate

It takes a force to make a mass accelerate.
The greater the mass, and the greater the acceleration, the more force is needed.

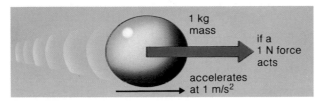

A 1 N force is needed to accelerate 1 kg at 1 m/s².
A 2 N force is needed to accelerate 1 kg at 2 m/s².
A 2 N force is needed to accelerate 2 kg at 1 m/s².
A 8 N force is needed to accelerate 4 kg at 2 m/s².

In all cases,

Force = mass × acceleration

In symbols,

$F = ma$

This is **Newton's second law of motion**.
This equation can be rearranged in two forms:

$$a = \frac{F}{m} \qquad m = \frac{F}{a}$$

You would use the left-hand equation, for example, if you knew the mass and the force, but needed to calculate the acceleration.

1 Copy and complete:

A __ is a push or pull. It is measured in __, or __ for short. Small forces can be measured with a __ __. All things resist acceleration. The effect is called __. The greater the __, the greater the resistance to acceleration. To give a __ of 2 kilograms an __ of 3 m/s², a force of __ is needed. Twice the __ pushing on the same __ would produce __ the acceleration.

2

What force is needed to make the rock accelerate at:
a 2 m/s² **b** 0.5 m/s² **c** 4 m/s²?

3 In an experiment with a 0.5 kg trolley, someone measures the pulling force, calculates the acceleration, writes down the values '10' and '5', but forgets to note which is which.
Can you decide, and add the correct units?

4

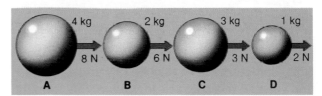

a Which masses have the same acceleration?
b Which mass has most acceleration?
c Which mass has least acceleration?

5 The three vehicles below are among the most powerful of their type.
Which wins the contest for the best acceleration? Which is the loser?

Boeing 747:
mass 400 000 kg
force from engines 800 000 N

Porsche 911:
mass 1300 kg
force from engine 7800 N

Honda 1000
mass 300 kg
force from engine 3000N

1.10 Weight: the pull of gravity

Does gravity always pull things downwards? Not according to cartoonists. They often use 'plausible impossibles' – things which seem reasonable, but aren't possible because they break the laws of physics. Walking off a cliff is probably the most well known. The character doesn't fall until he realises that he isn't standing on anything.

Gravitational force

Hang something from the end of a spring balance and you can measure the downward pull from the Earth. The pull is called a **gravitational force**.

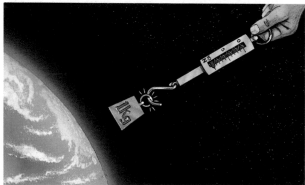

No one knows what causes gravitational force. But several things are known about it:
All masses attract each other.
The greater the masses, the stronger the pull.
The closer the masses, the stronger the pull.

The pull between small masses is far too weak to measure – less than one millionth of a newton between you and the person next to you for example. But the Earth has such a huge mass that the gravitational pull is strong enough to hold most things firmly on the ground.

Weight

Weight is another name for the gravitational force from the Earth.
As weight is a force, its unit is the newton.
On Earth, each kilogram of matter weighs 10 newtons.

People often use the word 'weight' when they really mean 'mass'. The person in the diagram doesn't 'weigh' 50 kilograms. He has a *mass* of 50 kilograms and a *weight* of 500 newtons.

Is there a link?

On Earth, the acceleration of free fall is 10 m/s².

On Earth, there is a gravitational force of 10 newtons on every kilogram.

These two facts are connected.
Try using the equation $F = m \times a$ to work out the acceleration of the two masses below.

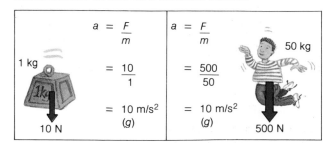

In each case, the acceleration is the same; *g*.
So you can think of *g*:
– as an acceleration of 10 m/s².
– as a **gravitational field strength** of 10 N/kg.

How to lose weight quickly

Go to the Moon. Even better, go deep into space, far away from all planets.

As different planets have different masses and sizes, your weight would vary from one place in the Universe to another. Take the case of a person with a mass of 50 kg:

	mass	weight
deep in space	50 kg	zero
near surface of the Moon	50 kg	80 N
near surface of Jupiter	50 kg	2700 N
near a black hole – a collapsed star whose gravitational pull is so great that even light cannot escape	50 kg	100 million million N

Travelling around space isn't going to get rid of the kilograms. Weight may vary, but mass stays the same. On the Moon for example, the gravitational pull is much less than on Earth. But the amount of matter in something is just the same. And it is just as difficult to speed up or slow down.

1 Copy the following, and fill in the blanks:
a __ is another name for the gravitational __ on something.
b Weight is measured in __.
c All masses __ each other. The closer the masses, the __ the pull between them.
2 Write down the weights of the following masses on Earth:

2 kg   4 kg   0.5 kg

3 'A bag of sugar weighs one kilogram.'
People might say this in everyday language. But the statement is wrong.
Why is it wrong?
What should it say?
4

A	B
$g = 10 \text{ m/s}^2$	$g = 10 \text{ N/kg}$

Describe in words what A and B tell you about a mass of one kilogram.
5 Aliens land on several planets, including Earth. Here is some information about the aliens:

alien	mass in kg	weight in N
A	40	80
B	20	200
C	10	200
D	20	40

a Which alien landed on Earth?
b Which two aliens landed on the same planet?
c The aliens have to jump from their spacecraft when they land.
Which alien will fall with the greatest acceleration?
d If all the aliens came to Earth, which would weigh least?
6 A 16 kilogram lump of rock weighs 10 N on the Moon.
a What is the acceleration of free fall on the Moon?
b What is the gravitational field strength on the Moon?
c How much would the rock weigh on Earth?

1.11 Balanced forces

The spacecraft Pioneer 10 was launched more than fifteen years ago. Now deep in space, it doesn't need engines to keep it moving. With no forces to slow it, it will keep moving for ever.

Sir Isaac Newton was the first to describe how things would move if no forces were acting on them. His **first law of motion** states:

If something has no force on it, it will:

  if still, stay still;

  if moving, keep moving at a steady speed in a straight line.

Slowed by friction

On Earth, unpowered vehicles quickly come to rest – slowed by the force of friction. Friction is the force that tries to stop materials sliding across each other. There is friction between your hands when you rub them together, and friction between your shoes and the ground when anything slides over it. Friction is partly due to tiny bumps on the surfaces, and partly due to atoms in the two materials which tend to stick to each other.

Fluid friction

Liquids and gases are called fluids. They can also cause friction. When a car is travelling fast on a motorway, air resistance is by far the largest frictional force pulling against it.

Using friction

Friction is needed to give your shoes grip on the ground.

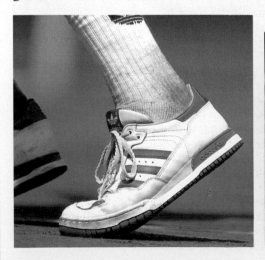

Reducing friction

Car bodies are designed so that the air flow is as smooth as possible. Less air resistance means less wasted fuel.

Forces in balance

Most things have several forces acting on them. For example, everything feels the pull of gravity, and moving things have friction trying to slow them down. Sometimes, the forces on something all cancel each other out. Then it behaves as if it has no force on it at all, and obeys Newton's first law.

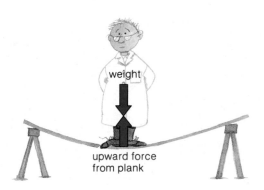

weight

upward force from plank

When the man stands on the plank, it sags, until the springiness of the wood produces enough upward force to oppose his weight. Then the forces cancel, so the man stays still. The ground isn't as springy as the plank, but it too produces an upward force to equal your weight when you stand on it.

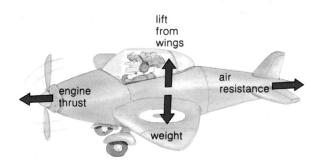

lift from wings

air resistance

engine thrust

weight

The aircraft is moving through the air at high speed. The weight of the aircraft is balanced by the lift from the wings. And air resistance is balanced by the thrust from the engine. The forces all cancel, so the aircraft keeps moving at a steady speed in a straight line.

If the thrust were *greater* than the air resistance, the aircraft would *gain* speed, and not keep a steady speed.

Terminal speed

As a skydiver falls, the air resistance increases as her speed rises. Eventually, air resistance is enough to balance her weight. If she weighs 500 newtons, then the air resistance rises to 500 newtons. She stops accelerating and falls with a maximum speed called her **terminal speed**. This is usually about 60 metres/second, though the actual value depends on air conditions, as well as her size, shape and weight.

If air resistance balances the skydiver's weight, why doesn't she stay still? There wouldn't be any air resistance unless she was moving.

Surely her weight is greater than air resistance if she is travelling downwards? No, if it were, she would gain speed, instead of keeping a steady speed.

Assume $g = 10\,\text{N/kg}$

1 If something has no forces acting on it at all, what happens to it if it is:
a still **b** moving?
2 Say whether friction is USEFUL or a NUISANCE in each of the following cases:
a a car tyre on the road;
b sledge runners on snow;
c a ship moving through the water;
d brake blocks on a cycle wheel;
e shoes on a pavement;
f a skydiver falling through air;
g someone's hand holding a screwdriver;
h a wheel spinning on an axle.
Which of these are examples of fluid friction?
3 A skydiver weighing $600\,\text{N}$ falls through the air at a steady speed of $50\,\text{m/s}$.
a Draw the diver, showing the forces acting on him.
b What name is given to his steady speed?
c What is the air resistance on him, in newtons?
d What is his mass?
e Why does he lose speed if he opens his parachute?

1.12 Action and reaction

Here are some pairs of forces:

In fact, no force can exist by itself.
All forces are pushes or pulls between *two* things.
So they *always* occur in pairs.
One force acts on one thing. Its equal but opposite partner acts on the other.

Sir Isaac Newton was the first to realise that forces occur in pairs. His **third law of motion** states:

For every action there is an equal and opposite reaction.
or

When A pushes on B, B pushes with an equal but opposite force on A.

If forces always occur in pairs, why don't they cancel each other out?
The forces in each pair are acting on two *different* things, *not* the same thing.

Why doesn't the ground move backwards when someone runs forward?
It does. But Earth is so massive, that the force has too small an effect to be noticed.

Rocks, rockets, and jets – Pairs of forces at work

Moving about in space is no problem if you have a rock handy.

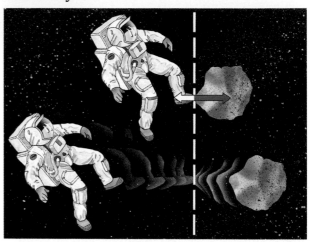

A pair of forces is produced when the astronaut pushes on the rock with her foot.
One force pushes the astronaut to the left.
An equal force pushes the rock to the right.
The astronaut has less mass than the rock. So the force has a greater effect on her.
She accelerates more than the rock.

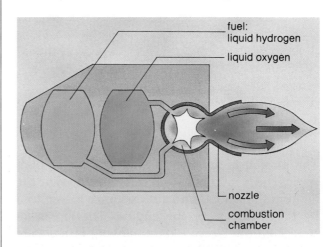

Rocket engines use a similar idea. But they push out large masses of gas, rather than rocks.
In a rocket engine, fuel and liquid oxygen are mixed together in the combustion chamber. The

fuel burns fiercely in the oxygen, turns to gas, and expands.

Huge forces are produced, which push engine and burning fuel apart.

One force pushes the burning fuel backwards. An equal but opposite force pushes the rocket forward.

How can a rocket accelerate through space if it has nothing to push against?

It does have something to push against – a huge mass of burning fuel. Fuel and liquid oxygen make up over 90% of the mass of a rocket.

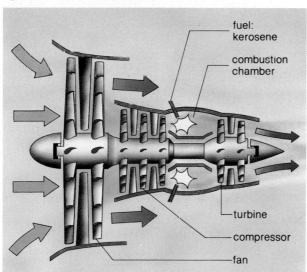

fuel: kerosene

combustion chamber

turbine

compressor

fan

A jet engine doesn't need a supply of liquid oxygen.

Instead, it uses a series of fans, called a compressor, to draw in large masses of air from the atmosphere.

The compressor is driven by the turbine.

The turbine is rather like a windmill. It is blown round by the fast-moving gases leaving the engine.

Most of the air drawn in by the jet engine doesn't go through the combustion chambers. It is pushed straight out by the huge fan at the front. This means less noise and better fuel consumption.

A large fan can push out over a quarter of a tonne of air every second.

1 One force in each pair is missing in the diagrams below. Copy the diagrams. Draw in and label the missing forces.

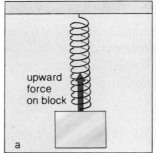

upward force on block

a

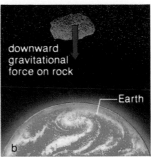

downward gravitational force on rock

Earth

b

2 Use the first letters of the answers below to make a word. You'll find this whenever there is any action.

a | ? | Engine using fuel and liquid oxygen.

b | ? | Every force has its __.

c | ? | If there is a reaction, it has to be there.

d | ? | It pushes air into the combustion chambers.

e | ? | This double-page is about the __ law of motion.

f | ? | First name of answer **h**.

g | ? | Forces in every pair are equal but __.

h | ? | He put forward the laws of motion.

3 Jet engines and rocket engines both push out large masses of gas.

Why can't a jet engine work in space?

Why aren't airliners powered by rocket engines?

4 Look at the diagram of the astronaut pushing on the rock. If the astronaut had more mass than the rock, what effect would this have?

5

A woman stands on a plank. Her weight is exactly balanced by an upward force from the plank.

Dave says that this is an example of Newton's third law of motion.

Sue says that it isn't.

Who is right? And why?

1.13 Momentum

On the right, you can see a massive truck travelling very fast. People sometimes say that a truck like this has lots of **momentum**. However, to scientists, momentum is something that can be calculated. They use this equation:

momentum = mass × velocity

For example if a model car of mass 2 kg has a velocity of 3 m/s:

momentum = 2 × 3 = 6 kg m/s

Like velocity, momentum is a vector (see page 12), so a + or − is often used to show its direction. For example:

if car moves to right: momentum = +6 kg m/s
if car moves to left: momentum = −6 kg m/s

Force and momentum

There is a link between force and momentum:

$$\text{Force} = \frac{\text{change in momentum}}{\text{time taken}}$$

You could use this equation to calculate the force needed to make a 2 kg model car accelerate from 3 m/s to 9 m/s in 4 seconds:

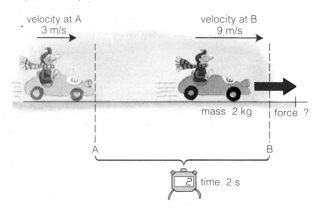

momentum (at 3 m/s) = 2 × 3 = 6 kg m/s
momentum (at 9 m/s) = 2 × 9 = 18 kg m/s
So, change in momentum = 18 − 6 = 12 kg m/s

So, force = $\dfrac{\text{change in momentum}}{\text{time taken}} = \dfrac{12}{4} = 3$

So, force needed = 3 newtons

Equation links

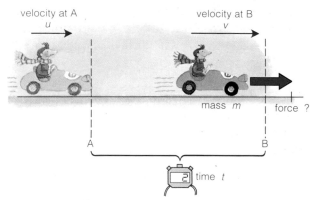

Above, a force makes a car of mass m accelerate from velocity u to velocity v in time t.

$$\text{Force} = \frac{\text{change in momentum}}{\text{time taken}} = \frac{mv - mu}{t} \quad (1)$$

$$= m\left(\frac{v-u}{t}\right)$$

But $\left(\dfrac{v-u}{t}\right)$ is the car's acceleration

$$\text{So, force = mass × acceleration} \quad (2)$$

This shows that (1) and (2) are really different versions of the same equation. This equation is sometimes called **Newton's second law of motion**.

Conserving momentum

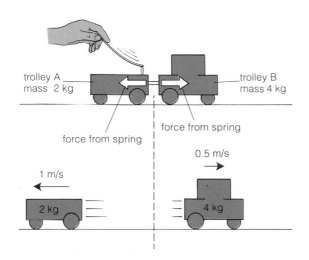

trolley A mass 2 kg

trolley B mass 4 kg

force from spring

force from spring

1 m/s

2 kg

0.5 m/s

4 kg

At first, the trolleys A and B above are at rest. Then a spring is released between them so that they shoot off in opposite directions. The trolley with the lower mass gains the higher velocity.

The diagram shows some typical mass and velocity values. Using these, you can work out the momentum changes taking place (note: a + or − is used to show motion to the right or left):

Before spring is released:
total momentum of trolleys = 0

After spring is released:
momentum of B = $4 \times (+0.5) = +2$ kg m/s
momentum of A = $2 \times (-1) = -2$ kg m/s

So, total momentum of trolleys = $2 + (-2) = 0$

Note that the total momentum *after* the spring is released is exactly the same as it was *before*. This is an example of the **law of conservation of momentum:**

Things may push or pull on each other but, if there is no force on them from outside, their total momentum stays the same.

You may be able to see why this law applies to the trolleys. The forces on A and B are equal but opposite. They act for the same time. So they produce equal but opposite changes in momentum. The change to the left cancels the change to the right, so the total momentum is unchanged.

Rocket thrust

Engineers can use the equation linking force and momentum to calculate the **thrust** (force) from a rocket or jet engine. For example:

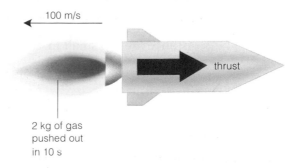

100 m/s

thrust

2 kg of gas pushed out in 10 s

In 10 seconds, a model rocket pushes out 2 kg of exhaust gas at a velocity of 100 m/s.

Change of momentum of exhaust gas
$$= 2 \times 100 = 200 \text{ kg m/s}$$

$$\text{Force} = \frac{\text{change in momentum}}{\text{time}} = \frac{200}{10} = 20 \text{ newtons}$$

This force pushes the exhaust gas backwards. By Newton's third law, an equal but opposite force must be pushing the rocket forward. So:

thrust from rocket engine = 20 newtons

	mass in kg	velocity in m/s
car	1000	5
motor cycle	200	30

1 In the table above **a** What is the momentum of the car? **b** Which has more momentum, the car or the motor cycle? **c** Which would have more momentum if they were both travelling at the same velocity?
2 What force is needed to make a 10 kg mass accelerate from 2 m/s to 4 m/s in 5 seconds?
3 What is the force from a rocket engine which, in 2 seconds, pushes out 4 kg of exhaust gas at a velocity of 200 m/s?
4 Look at the astronaut on page 28. If the astronaut has a mass of 100 kg, the rock has a mass of 200 kg, and the rock gains a velocity of 2 m/s to the right, what velocity does the astronaut gain to the left?

1.14 Curves and circles

Downwards, sideways ...

Below, you can see what happens if one ball is dropped and another is thrown sideways at the same time. (Here, the balls are heavy and the effects of air resistance are too small to notice.)

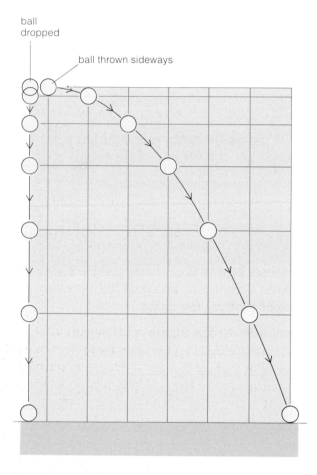

ball dropped

ball thrown sideways

The diagram shows the positions of the balls at regular time intervals (every 1/10 second). If you study the diagram, you will notice two things:

- Both balls hit the ground at the same time. They have exactly the same downward acceleration (g, 10 m/s²).

- As it falls, the second ball moves sideways over the ground at a steady speed. In other words, its horizontal velocity is constant.

Results like this show that, if something is falling freely, its vertical and horizontal movements are quite independent of each other.

... and into orbit

The diagram below shows a 'thought experiment'. An astronaut is standing on a tall tower, high above the atmosphere, where there is no air resistance. She is so strong that she can throw a ball at the speed of a rocket!

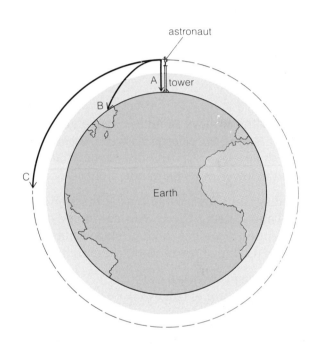

astronaut

A tower

B

C

Earth

Ball A is dropped. It accelerates downwards, straight to the ground.

Ball B is thrown horizontally. It too accelerates downwards. But it also moves sideways at a steady speed.

Ball C is thrown horizontally, but much faster. Once again, the ball is falling. But this time, its sideways speed is so fast that the curve of the fall matches the curve of the Earth. The ball is accelerating downwards, but getting no closer to the ground! It is in **orbit** around the Earth.

Satellites are put in orbit by fast, powerful rockets. For a near-Earth orbit, the speed required is about 8 km/s (18 000 mph). In orbit, a satellite is in free fall, just like a ball.

Moving in circles

Many orbits are circular. Below, you can see another example of circular motion: a ball being whirled round on the end of a piece of string. The person in the middle keeps the ball moving at a steady speed.

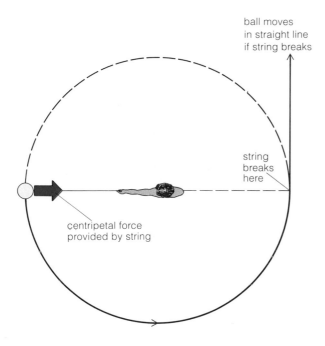

An inward force is needed to make the ball move in a circle. This is called the **centripetal force**. Without it, the ball would move in a straight line, as predicted by Newton's first law of motion.

With the ball above, the string provides the centripetal force. *More* force is needed:
- if the mass of the ball is *greater*
- if the speed of the ball is *greater*
- if the radius of the circle is *less*.

1 Below, one bullet is shot from a gun at 50 m/s while another is dropped. The dropped bullet takes 2 seconds to reach the sea. There is no air resistance.

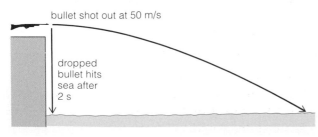

For a satellite orbiting the Earth, gravity provides the centripetal force needed to make it move in a circle.

For a car cornering at high speed, sideways friction from the tyres provides the centripetal force needed to make it turn the corner.

Centrifugal or centripetal?

When you whirl a ball round on a piece of string, you may say that you feel an outward 'centrifugal force'. But there is no outward force on the ball. If the string breaks, the ball moves off in a straight line, at a tangent. It isn't flung outwards.

Moving a ball in a circle doesn't *produce* a centripetal force. If the centripetal force is, say, 20 newtons, this tells you that a 20 newton force is needed to make the ball move in the circle.

a How long does the bullet from the gun take to reach the sea?
b What is the horizontal velocity of this bullet when it reaches the sea?
c How far is it from the cliff when it reaches the sea?
2 A car goes round a corner:
a What provides the centripetal force?
b Is *more* force or *less* force needed if the car is
i heavier ii faster iii going round a tighter curve?
c What would happen to the car if there were no centripetal force?

1.15 Speedy delivery

Out of the way!

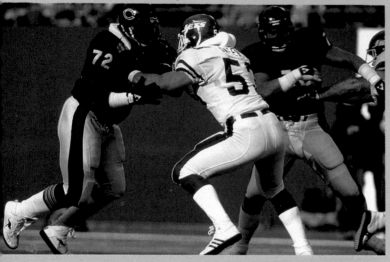

William 'The Refrigerator' Perry, lead blocker for the Chicago Bears. His job is to charge through the defence lines, clearing gaps for the running backs to pass through. Mass is the secret of his success. At 145 kilograms, he has more than twice the mass of the average male. And once he starts running he is extremely difficult to stop.

Someone else who needs mass as well as strength. When she pushes the shot forward, there's a backward push on her body which slows her down and reduces the speed of the shot. The more mass she has, the less effect this backward push has.

Who wins over 1500 metres?

1.5 m/s   7 m/s   12 m/s   15 m/s

30 mph

Swimmers lose out because water resistance is much higher than air resistance. Cyclists are fastest of all. And even faster if there's another vehicle in front to reduce the air resistance. The

highest speed ever reached on a bicycle was 63 metres/second (140 mph) – behind a car with a windshield on the back.

Fast service

Nice style. Shame about the action.
Could you work out the speed of the racket from this photograph? If not, why not?

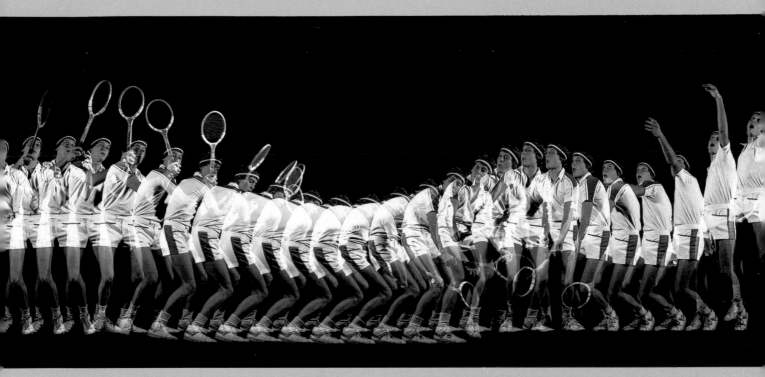

Getting the elbow

A top player uses a racket with tight strings to help her serve the ball really fast. She tries to hit the ball near the centre of the racket. Otherwise, the forces can injure her elbow.
An ordinary player uses racket strings that are less tight. The strings stretch more when the ball is hit, which cuts the speed down. But, if the player hits the ball off-centre, the forces on the elbow are small and not so damaging.

Explain why:
- runners can travel faster than swimmers;
- cyclists can travel faster than runners.

In Olympic speed events, competitors need to keep their air resistance as low as possible. Try to find out how the following reduce their air resistance:
- sprint cyclists;
- speed skaters;
- swimmers.

Describe what happens to the speed of a tennis racket from the beginning of a serve to the end.

Make a list of the games players or athletes for whom plenty of mass is:
- an advantage;
- a disadvantage.

1.16 Turning effects

What's needed to turn these?

Something which increases the turning effect of your hand.

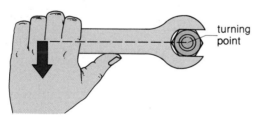

The nut is easy to turn with a spanner.

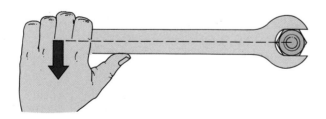

It is easier still if the spanner has a long handle. You can increase the turning effect in two ways:
1 Increase the force.
2 Move the force further away from the turning point of the nut.

Moments

The turning effect of a force is called a **moment**. It can be calculated as follows:

Moment = force × distance from turning point

Moments are **clockwise** or **anticlockwise** depending on which way they turn. For example:

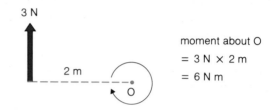

This force has an anticlockwise moment of 6 N m point O.

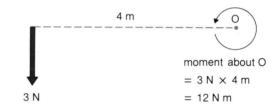

This force has an anticlockwise moment of 12 Nm about point O. It has twice the turning effect, but in the opposite direction.

Torque

In engines and motors, several forces act together to produce a turning effect. The turning effect is called a **couple** or **torque**. Typical torque values are:

Moments in balance

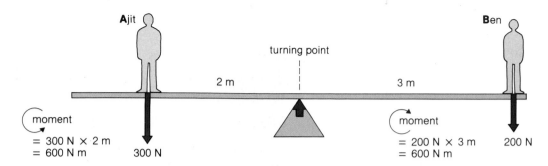

There are two turning effects at work on the sea-saw; Ajit has an **anticlockwise** turning effect; Ben has a **clockwise** turning effect. The two moments are equal. So their turning effects cancel, and the see-saw balances. This is an example of the **principle of moments**:

If something is in balance, about the turning point the total clockwise moment is equal to the **total anticlockwise moment.**

The principle works in more complicated cases as well. In the diagram below, there is one anticlockwise moment about the turning point, but two clockwise moments. Add up the two clockwise moments.

The total is the same as the anticlockwise moment. So the see-saw balances.

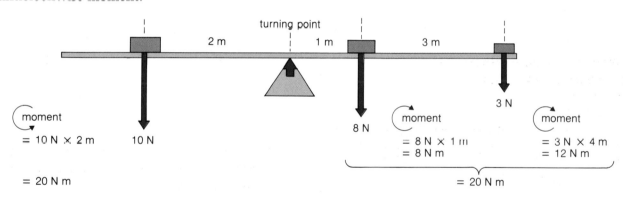

1

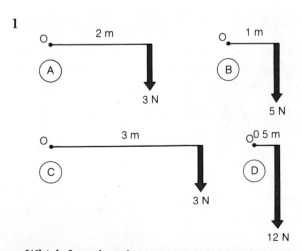

a Which force has the greatest moment about O?
b Which forces have the same moment about O?
c Which force has the least moment about O?

2

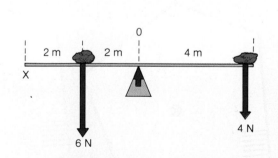

Someone is trying to balance a plank with stones.
a Calculate the moment of the 4 N force about O.
b Calculate the moment of the 6 N force about O.
c Will the plank balance?
If not, which way will it tip?
d What extra force would be needed at X to balance the plank?

37

1.17 Centre of gravity

Balancing for Gold

Most people would find it impossible to walk along such a narrow beam, let alone perform handstands and somersaults on it.

The secret lies in how you position your weight. Every particle in your body has a small gravitational force acting on it. Together, these forces act like a single force pulling at just one point. This single force is your **weight**.

The point is called your **centre of gravity** or **centre of mass**.

Keep your centre of gravity over the beam and you stay on. Allow it to move to one side, and your weight produces a turning effect which tips you off.

Simple shapes, like a metre rule, often have a centre of gravity exactly in the middle. Vehicles usually have a low centre of gravity because most of their heavy mechanical parts are low down.

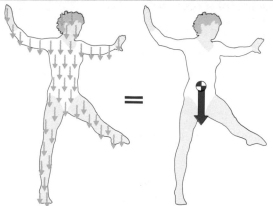

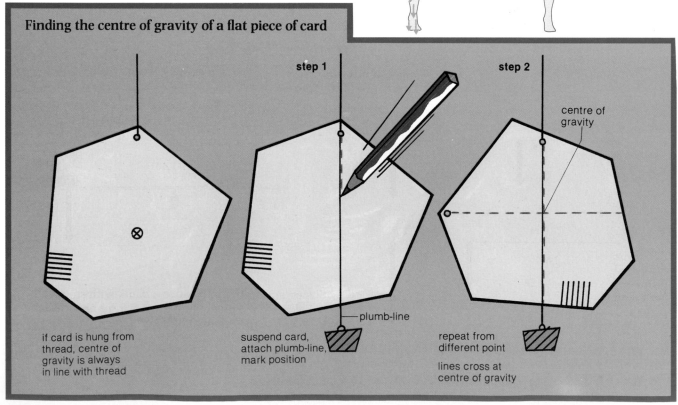

Finding the centre of gravity of a flat piece of card

if card is hung from thread, centre of gravity is always in line with thread

step 1

suspend card, attach plumb-line, mark position

plumb-line

step 2

centre of gravity

repeat from different point

lines cross at centre of gravity

How stable?

base

base

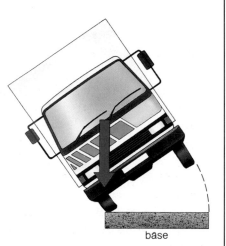

base

If something won't topple over, its position is stable:

This truck is in a stable position. If it starts to tip, its weight will pull it back again. As long as its centre of gravity stays above its base, it won't topple over.

This racing car is even more stable than the truck. It has a lower centre of gravity and a wider base. It could be tipped much further before it started to topple.

Clever stunt driving – but it has put the truck in an unstable position. If the truck tips any further, its centre of gravity will pass over the edge of its base. Then its weight will pull it right over.

Equilibrium

Like the vehicles above, the shapes below are all in a state of balance or **equilibrium**.

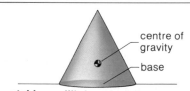

centre of gravity
base

stable equilibrium
Even if you tip the cone a little, the centre of gravity stays over the base.

unstable equilibrium
The cone is balanced, but not for long. Its pointed 'base' is so small that the centre of gravity immediately passes beyond it.

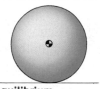

neutral equilibrium
Left alone, the ball stays where it is. Moved, it stays in its new position. Wherever the ball lies, the centre of gravity is always exactly over the point where it touches the bench.

1 In the diagram, the kitchen stool is about to topple over. Copy the diagram and mark on the position of the centre of gravity.

Would the stool be MORE stable, or LESS stable, if it had:
a a higher centre of gravity?
b a wider base?
Explain why three-legged stools aren't as stable as stools with four legs.

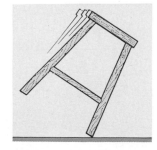

2 a Redraw the diagram, showing the weight as a force arrow.

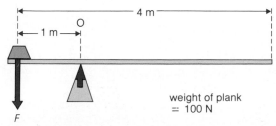

4 m

O

1 m

weight of plank = 100 N

F

b How far is this force from the point O?
c What is the moment of this force about O?
d If the plank balances, what must be the moment of the force F about O?
e What is the value of F?

1.18 Stretching and compressing

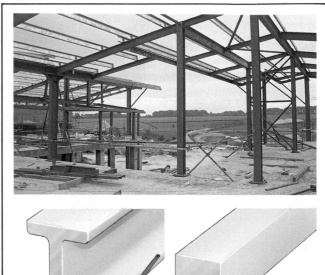

At 555 metres, the CN tower in Toronto is the tallest building in the world. However, its steel and concrete structure isn't quite as rigid as it looks. In high winds, the top can sway up to half a metre. And the tower is actually shortened by several centimetres because the structure is compressed by its own weight.

Whenever several forces act on something, its shape changes – though sometimes only by a small amount.

Some things are designed to bend and twist. Some springs for example. However, the steel frames used in most modern buildings are designed to change shape as little as possible. The frames are made using I-section beams. In every case, the more cross-sectional ('end-on') area the bar has, the more it resists being put out of shape.

An I-section beam has a much greater resistance to bending than a solid square section beam made from the same amount of metal.

Acting together, forces can have the following effects:

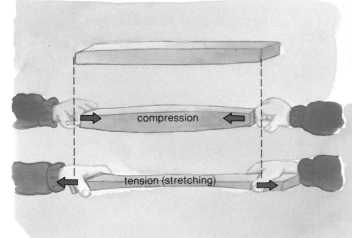

Elastic and plastic

To engineers, these don't have quite the same meaning as in everyday language.
Bend a ruler a little. Then release it.
It goes back to its original shape.
Materials which behave in this way are **elastic**.

Press a piece of Plasticine, then release it. It doesn't return to its original shape.
Materials which behave in this way are **plastic**.

A bumper on a car is elastic – provided it isn't bent too far. Given too much force, it passes its **elastic limit** and stays out of shape.

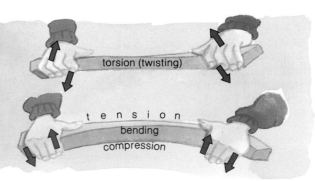

Hooke's law

Many materials obey a simple law when compressed or stretched. Take the case of a spring:

The spring is stretched in stages by hanging masses from one end. The stretching force is called the **load**.
As g = 10 N/kg, there is a 1 N load for every 100 g hung from the spring.
Each time the load is changed, the **extension** of the spring is measured. The extension is the difference between the stretched length of the spring and its original unstretched length.
Typical readings are shown in the chart.

The readings can be used to plot a graph of extension against load.
Up to the point E:
1 The graph is a straight line through O.
2 Every extra 1 N of load produces the same extra extension (10 mm in this case).
3 If the load is doubled, the extension is doubled, **and so on.**
Mathematically, these all mean that **the extension is directly proportional to the load.** This is sometimes called **Hooke's law.**

E is the **elastic limit** of the spring. If this point is passed, the spring doesn't go back to its original length when the load is removed. It ends up longer than before.

Steel bars don't stretch as much as springs. But they obey the same law. So do glass, wood, and many other materials.

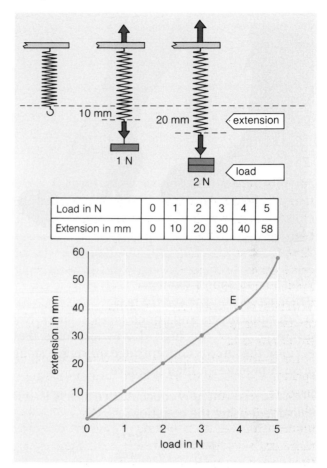

Load in N	0	1	2	3	4	5
Extension in mm	0	10	20	30	40	58

1 a Write down the parts of the ridge tent which are under TENSION; in COMPRESSION; BENDING.

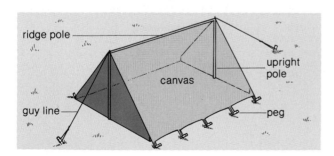

b What would happen to the ridge pole if its ELASTIC LIMIT were exceeded?
c The force on the guy line is increased. The guy line doesn't break. What does happen to it?
How would the result be different if a thicker guy line, made from the same material, were used?

2 The table shows the readings taken in a spring-stretching experiment:

Load in N	0	1	2	3	4	5	6
Length in mm	40	49	58	67	76	88	110
Extension in mm							

a Copy and complete the table.
b What is the unstretched length of the spring?
c Plot a graph of EXTENSION against LOAD.
d Mark the ELASTIC LIMIT on your graph. What happens to the spring beyond this point?
e What load is needed to produce an extension of 35 mm?
f What load is needed to make the spring stretch to 65 mm long?

1.19 Pressure

Which causes most damage?
Believe it or not, the stiletto heel.
It can ruin carpets and punch holes in floors. Not just because of the high downward force. But because the force is concentrated on such a small area. It produces a high **pressure**.

Pressure tells you how concentrated a force is. It is calculated using the equation:

$$\text{pressure} = \frac{\text{force}}{\text{area}}$$

The unit of pressure is the **newton/metre² (N/m²)** or **pascal (Pa)**.

What the figures mean

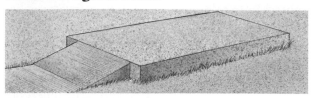

Pressure under concrete garage base: 8000 N/m².
There is a force of 8000 newtons on every square metre of ground.

Pressure under stiletto heel: 2 000 000 N/m².
Much less than a square metre here of course. But the heel has the same squashing effect on the floor as a 2 000 000 newton force spread over a full square metre.

Pressure problems

Rearrange the pressure equation, and you get:

$$\text{force} = \text{pressure} \times \text{area}$$

This equation is useful if you know the pressure, and the area over which it is acting, but you need to find the force.

For example:

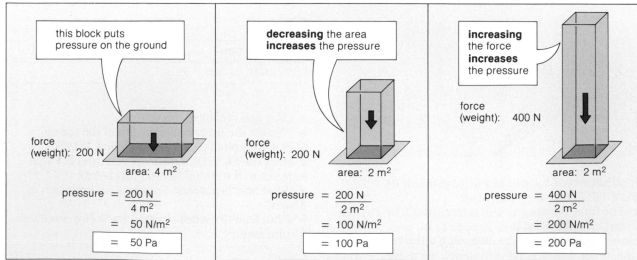

this block puts pressure on the ground

force (weight): 200 N

area: 4 m²

$$\text{pressure} = \frac{200\ \text{N}}{4\ \text{m}^2}$$

$$= 50\ \text{N/m}^2$$

$$= 50\ \text{Pa}$$

decreasing the area **increases** the pressure

force (weight): 200 N

area: 2 m²

$$\text{pressure} = \frac{200\ \text{N}}{2\ \text{m}^2}$$

$$= 100\ \text{N/m}^2$$

$$= 100\ \text{Pa}$$

increasing the force **increases** the pressure

force (weight): 400 N

area: 2 m²

$$\text{pressure} = \frac{400\ \text{N}}{2\ \text{m}^2}$$

$$= 200\ \text{N/m}^2$$

$$= 200\ \text{Pa}$$

keeping the **area** high and the **pressure** low	keeping the **area** low and the **pressure** high
walking on sand hurts less than walking on pebbles: less pressure means less pain! 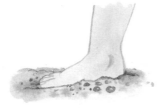	studs on hockey or football boots: enough pressure here for them to sink into the ground 
load-spreading washer ensures that the head of the bolt isn't pulled through the woodwork 	blade on knife: the sharper it is, the higher the pressure 
heavy animals need thick legs, or their bones wouldn't cope with the pressure 	point on drawing pin: far more pressure than wood can stand 

Assume $g = 10$ N/kg.

1 Copy out, and fill in the blanks:
Pressure tells you how concentrated a __ is. It is measured in __ or __, and is calculated using the equation: pressure = __ __. A force of 12 N acting over an area of $2\,m^2$ causes a pressure of __. If the area were less, the pressure would be __.

2 The suitcase, the hover mower and the paving stone are all resting on the ground.

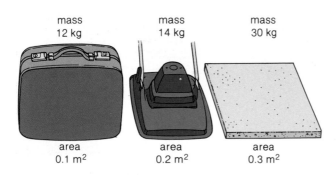

mass 12 kg	mass 14 kg	mass 30 kg
area 0.1 m²	area 0.2 m²	area 0.3 m²

Calculate:
a the weight of each;
b the pressure on the ground from each.

3 Sid is heavy. He weighs 720 N.
He also has big feet. They cover an area of $0.12\,m^2$.

0.12 m²

The ice can stand a pressure of 5000 Pa before it cracks. Sid thinks he is safe. Is he right?
4 Sid can't iron the creases out of his jeans unless the pressure on them is at least 1500 Pa.
a If Sid's iron has an area of $0.02\,m^2$, what downward force is needed to produce this pressure?
b If the iron weighs 10 N, what downward force must Sid use on the iron to get rid of the creases?

1.20 Pressure in liquids

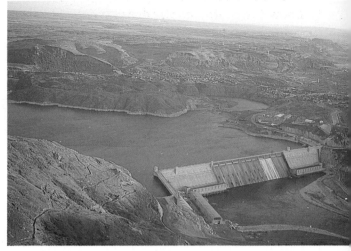

Grand Coulee Dam

It took nearly nine years and 20 million tonnes of concrete to build the Grand Coulee Dam in Washington State, USA. At its base, the dam is more than 60 metres thick. It has to be. Otherwise it would never withstand the pressure from the 150-metre-deep lake of water on the other side.

Gravity pulls any liquid downwards in its container. This puts pressure on the container. And it puts pressure on anything put into the liquid.

The pressure pushes in all directions A liquid under pressure pushes on every surface it touches, whether the surface is facing upwards, downwards or sideways.

The pressure increases with depth Dams are thicker at the base than the top because they have to withstand a greater pressure at the bottom of the lake. The deeper into a liquid you go, the greater the weight of liquid above.

The pressure is affected by the density of the liquid Put petrol in the lake instead of water, and the pressure everywhere would be less. Petrol is less dense than water, so there is less weight to produce the pressure.

The width or shape of the container doesn't affect the pressure In the diagram, the pressure at the bottom of each 'lake' is the same. The bottom of the wider lake does have a greater weight of water to support. However, that weight is spread over a larger area.

A submarine is built to withstand water pressure in all directions.

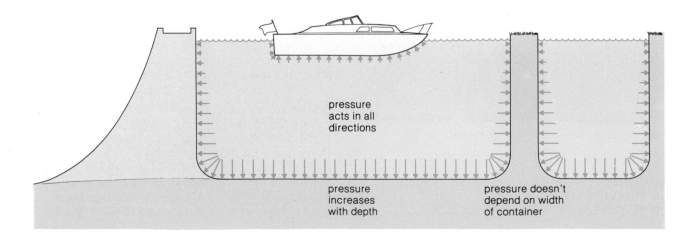

pressure acts in all directions

pressure increases with depth

pressure doesn't depend on width of container

Working out the pressure

You can calculate the pressure at any point in a liquid provided you know the depth and the density of the liquid:

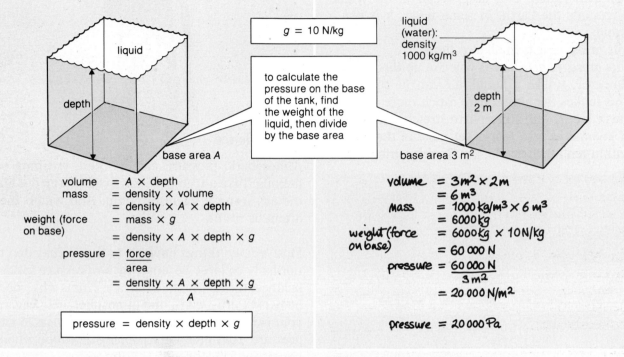

$g = 10$ N/kg

liquid
(water):
density
1000 kg/m³

to calculate the pressure on the base of the tank, find the weight of the liquid, then divide by the base area

depth
2 m

base area A

base area 3 m²

volume = A × depth
mass = density × volume
= density × A × depth
weight (force on base) = mass × g
= density × A × depth × g
pressure = force / area
= density × A × depth × g / A

pressure = density × depth × g

volume = 3m² × 2m
= 6m³
mass = 1000 kg/m³ × 6 m³
= 6000 kg
weight (force on base) = 6000 kg × 10 N/kg
= 60 000 N
pressure = 60 000 N / 3m²
= 20 000 N/m²

pressure = 20 000 Pa

The base area doesn't affect the final answer. The pressure is the same, no matter how wide or narrow the tank.

Assume $g = 10$ N/kg;
the density of water = 1000 kg/m³.

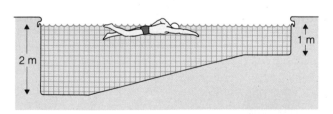

Calculate the water pressure at the bottom of this swimming pool:
a at the shallow end   **b** at the deep end.

2 Calculate the water pressure on a scuba diver as she descends into a freshwater lake at:
a 10 m below the surface;
b 20 m;
c 30 m.
d What effects do you think that this pressure will have on her?

3 Sian's shower won't work properly because the pressure at the shower-head end is too low. She decides to solve the problem by lengthening the pipe and raising the tank.

0.75 m

What is the water pressure at the shower-head? How high must she raise the tank so that the pressure at the shower-head is 15 000 Pa?
4 What is the water pressure on the submarine in the photograph opposite if it dives to a depth of 50 m in fresh water?
How deep can the submarine safely dive in fresh water if its hull can withstand a pressure of 1 000 000 Pa?
Sea water is more dense than fresh water. Is the safe diving depth in sea water MORE or LESS than in fresh water?

1.21 Pressure from the atmosphere

The atmosphere is a deep ocean of air which surrounds the Earth. In some ways, it is like a liquid:
– its pressure acts in all directions;
– its pressure gets less as you rise up through it.
However, unlike a liquid, air can be squashed. This makes the atmosphere much more dense at lower levels. The atmosphere stretches hundreds of kilometres into space. Yet most of the air lies within ten kilometres of the Earth's surface:

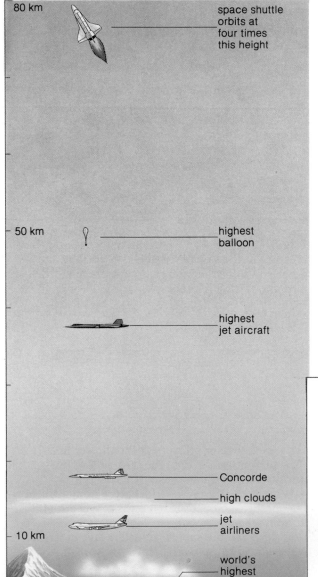

The evidence

Atmospheric pressure can be quite crushing – literally. Take an 'empty' metal can. Pump out all the air, using a vacuum pump. And watch the dramatic result.

Most 'empty' things have air in them. They don't normally collapse because the air pressure inside balances the air pressure outside. This is why your body isn't crushed by the atmosphere. However, your ears are very sensitive to small *changes* in air pressure. They give you a popping sensation when you rise quickly through the atmosphere – in a car travelling uphill, for example.

How much?

Down at sea-level, atmospheric pressure is about 100 000 Pa.
That's equivalent to the weight of ten cars pressing on every square metre. Or the pressure from a column of water 10 metres deep:

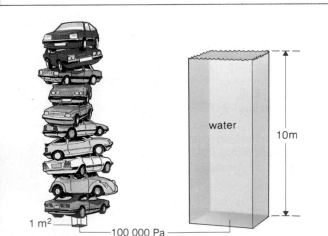

Using atmospheric pressure

Sucking through a straw

You use your lungs to lower
the air pressure in the straw.
Atmospheric pressure
does the rest.
It *pushes* the liquid
up the straw.

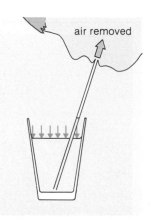

air removed

Vacuum cleaner

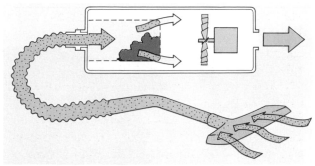

The fan lowers the air pressure just beyond the bag. The
atmosphere rushes in, carrying dust and dirt with it.
The bag stops the dust, but not the air.

Power-assisted car brakes

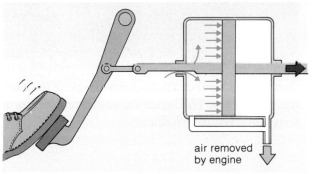

air removed
by engine

Extra help from the atmosphere when you push on the pedal.
Outside air enters the empty cylinder and pushes the piston to
the right.

Vacuum forming

Remove the air under a sheet of warm, soft plastic, and the
atmosphere will push the plastic into any shape you like.

Assume atmospheric pressure is 100 000 Pa – the
same as the pressure due to a column of water 10 m
deep.

Jeff works as a diver on an oil rig. He regularly dives
to the sea bed, 50 m beneath the surface, to inspect
and maintain the rig.
Jeff's breathing equipment supplies him with a
special air mixture. To prevent him being crushed,
the air pressure is automatically adjusted so that it
exactly balances the outside pressure.
What is the pressure of the air he breathes when he
is working:

a at sea-level?

b 10 m beneath the surface?

c on the sea-bed?

Jeff's breathing equipment can't function safely at
pressures above 800 000 Pa. Jeff's boss wants him to
take on a salvage operation at a depth of 100 m.
Should he refuse? How deep can he safely go?

2 Is it possible to 'suck' the water out of the bottle in
the diagram? If not, why not?

1.22 Measuring air pressure

Who needs to?

Weather forecasters know that atmospheric pressure changes slightly from hour to hour, depending on the weather. Rain clouds form in areas of lower pressure called depressions. So a fall in pressure often means that bad weather is on the way.

Skin-divers need to know the pressure of the air in their cylinders. The lower the pressure, the less air they have left to breathe.

Pressure measurers

Instruments which measure atmospheric pressure are called barometers.

Mercury barometer This is a long glass tube, sealed at one end, with the other end dipped in a dish of liquid mercury. The atmosphere presses on the mercury in the dish and this keeps a column of mercury up the tube. The greater the pressure, the higher the column. The space in the top of the tube is completely empty – it is a **vacuum**. If air were trapped in the tube, its pressure would stop the mercury rising so far.

Atmospheric pressure is sometimes measured in 'millimetres of mercury', rather than pascals. At sea-level, its average value is about 760 millimetres of mercury.

Mercury is used in a barometer, rather than any other liquid, because it is so dense. If the barometer contained water, the tube would have to be over 10 metres long.

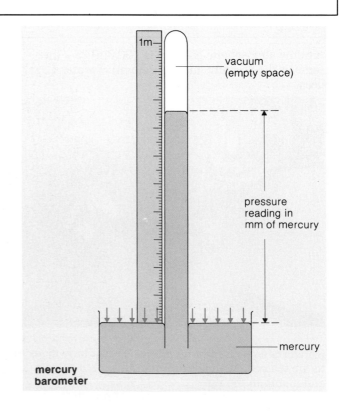

aneroid barometer

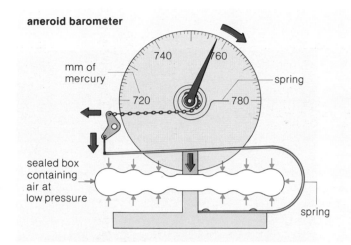

manometer

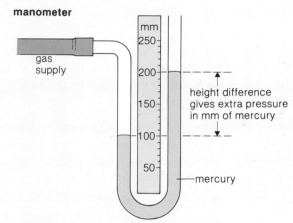

height difference gives extra pressure in mm of mercury

Aneroid barometers These are the barometers which people hang in the hall to forecast the weather. They aren't as accurate as a mercury barometers. But they are more robust, simpler to read, and easier to move about.

In the middle of the barometer is a sealed metal box with flexible sides. Most of the air has been removed from the box, so it is squashed by the atmosphere. The greater the pressure, the more the sides are pushed in, and the further the pointer moves round the scale.

Pilots and mountaineers use a special type of aneroid barometer called an **altimeter**. Its reading tells them their height above sea-level.

Manometers are used to measure gas and liquid pressures. The one in the diagram is filled with mercury. The height difference tells you the *extra* pressure that the gas supply has on top of atmospheric pressure – 100 mm of mercury in this case. If you want to know the actual pressure of the gas, you need to add on the value of atmospheric pressure.

Bourdon gauges are the pressure gauges you see on the top of gas cylinders. They aren't as accurate as manometers, but they are tougher and more portable. They contain a bent hollow metal tube which partly unbends when the pressure inside rises. This moves a pointer across a scale.

Assume $g = 10 \, \text{N/kg}$

1

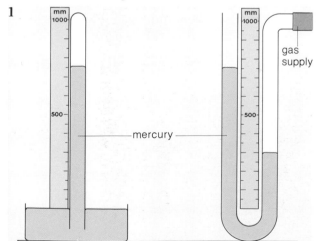

Study the diagram, then write down:
a the name of instrument on the left;
b the name of instrument on the right;

c the value of atmospheric pressure, in mm of mercury;
d the reading on instrument on the right;
e the pressure of the gas supply.
2 BOURDON GAUGE, MERCURY BAROMETER, ANEROID BAROMETER, MANOMETER
Say which of these instruments you would use to measure:
a atmospheric pressure, as accurately as possible;
b atmospheric pressure half-way up a **mountain**;
c the pressure of gas in a cylinder;
d the pressure of a laboratory gas supply, as accurately as possible.
3

| Mercury barometer: |
| Height of mercury column = 760 mm of mercury |
| Density of mercury = 13 600 kg/m³ |

Use the information above, and the equation on **page 45**, to calculate a value for atmospheric pressure in pascals.

1.23 Floating and sinking

The whale in the photograph is more than a hundred times heavier than you, yet it floats easily. Its huge weight is supported by the pressure of the water around it.

When something is in water, the water pressure pushes in on all sides. However the pressure underneath is more than the pressure above because of the greater depth. The result is an upward force called an **upthrust**. To feel an upthrust for yourself, try pushing an empty plastic container down into some water.

Archimedes' principle

Liquids and gases are called **fluids** ('fluid' means 'a substance which flows'). When something is put into a fluid, it takes up space once occupied by the fluid. Fluid is **displaced** (pushed out). Scientists have found that:

When something is in a fluid, the upthrust is equal to the weight of fluid displaced.

This is called **Archimedes' principle** (after the Greek scientist described on page 10). Note:

- Archimedes' principle is true for all liquids and gases, not just water. It applies whether something is completely immersed in a fluid, or only partly immersed.

- The bigger something is, the more fluid it will displace, so the greater the upthrust will be.

Here is an example of Archimedes' principle:

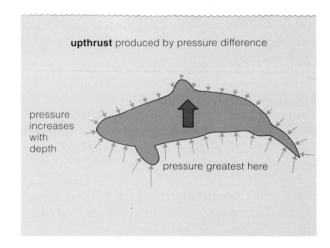

upthrust produced by pressure difference

pressure increases with depth

pressure greatest here

Floating

Something will float provided the upthrust is strong enough to support its weight. For example, this boat is floating in water:

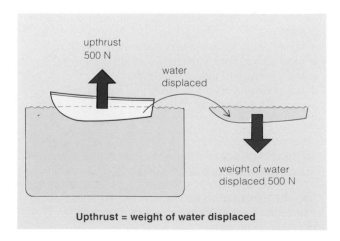

upthrust 500 N

water displaced

weight of water displaced 500 N

Upthrust = weight of water displaced

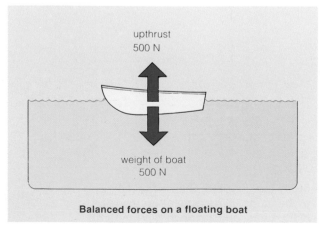

upthrust 500 N

weight of boat 500 N

Balanced forces on a floating boat

Look at the boat diagrams on the previous page:

- As the boat weighs 500 newtons, the upthrust must also be 500 newtons. The boat is not sinking, so the two forces must balance.
- As the upthrust is 500 newtons, the weight of water displaced must also be 500 newtons (from Archimedes' principle).

Note that the weight of water displaced is equal to the weight of the boat. This is an example of the **law of flotation:**

If something is floating, the weight of fluid displaced is equal to the weight of the thing which is floating.

As you load up a boat, it floats lower in the water. This means that it displaces more water, so there is a greater upthrust to support the extra weight. However, if you add too much weight, the upthrust cannot support it, and the boat sinks.

Floating and density

By comparing density values, you can decide whether something will float in a particular fluid or not. (Density = mass/volume : see page 8.)

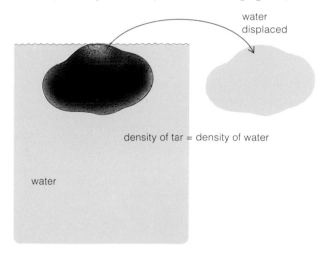

water displaced

density of tar = density of water

water

Tar floats in water, but only just. The law of flotation tells you that it has the same weight (and therefore the same mass) as the water it displaces. It also has the same volume. So it must have the same density. If its density were any greater, it would sink.

If a material floats, its density must be less than (or equal to) that of the fluid around it.

Floating in air

In large volumes, air is heavy. This hot-air balloon displaces more than 3 tonnes of air. It will float upwards if its total weight (including the air inside it) is less than the weight of air displaced.

Before take-off, cold air in the balloon is heated by gas burners to over $100\,°C$. This makes the air expand, so that over half a tonne of it is pushed out through the hole in the bottom. With less air to carry, the upthrust can support the fabric, burners, basket, and crew as well. So the balloon starts to rise.

1 A small boat weighs 1000 newtons. It is floating in water.
a Draw a diagram to show the forces acting on the boat.
b What is the upthrust on the boat?
c What weight of water is displaced by the boat?
d Extra weight is loaded into the boat. Explain why this will make the boat float lower in the water.
2 Explain why a 1 tonne boat will float in water, but a 1 tonne block of steel will sink.
3 A material has a density of $900\,kg/m^3$. Using information you can find on page 8, decide whether the material will float **a** in water; **b** in petrol. Explain each of your answers.

Questions on Section 1

1 Emma and Jane are having a race. The graph below shows how the speed of each athlete changes with time.

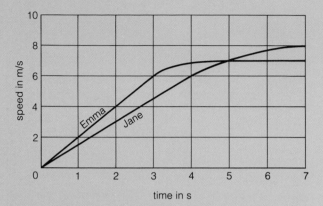

a Which athlete reaches the higher speed?
b Which athlete has the greater acceleration at the start of the race?
c After how many seconds does Jane's speed become greater than Emma's?
d What is Emma's maximum speed?
e To begin with, Emma's acceleration is $2\,m/s^2$. What does this figure tell you about the way her speed changes?
f What is Jane's acceleration over the first 4 seconds?
g What distance does Jane travel in the first 4 seconds?

2 A skydiver falls from a hovering helicopter. She waits a few seconds before opening her parachute. The table below shows how her speed changes with time from the moment she jumps:

time in s	0	1	2	3	4	5	6	7	8
speed in m/s	0		20	30	22	14	12	9	9

a Copy and complete the table, filling in the missing number.
b Plot a graph of speed against time.
c After how many seconds does the skydiver open her parachute? How can you tell from your graph?
d As the skydiver falls, there is a *downward* force acting on her and an *upward* force.
  i What causes the downward force?
  ii What causes the upward force?
  iii After 2 seconds, which of these two forces is the larger?
  iv After 8 seconds, how do the two forces compare?

e How would you expect your graph to be different if the skydiver's parachute were larger? (You could answer this by drawing sketches to show how the graph changes.)

3 In each of the following cases, decide whether the frictional forces should be as *low* as possible or as *high* as possible:
a Shoes in contact with the pavement.
b Brake blocks being pressed against the rim of a bicycle wheel.
c Hands holding the handlebars of a bicycle.
d Skis sliding over snow.
e Car tyres in contact with a road surface.
f A wheel turning on its axle.

4 The diagram below shows a model crane. The crane has a movable counterbalance.

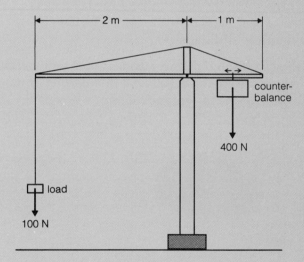

a Why does the crane need a counterbalance?
b Why does the counterbalance need to be movable?
c In the diagram, what is the moment of the 100 N force about O?
d To balance the crane, what moment must the 400 N force have?
e How far from O should the counterbalance be positioned?
f Where would you expect the counterbalance to be positioned if the crane is lifting its maximum load?
g What is the maxiumum load the crane should lift?
h Describe two ways of making the design of the crane more stable.

5 The diagram below shows how a steel spring stretches when a weight is hung from it.

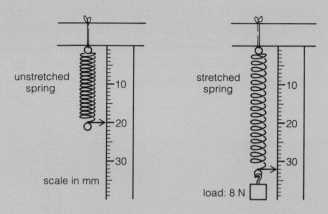

a What is the extension of the spring?
b If the spring obeys Hooke's law, what does this tell you about the way the extension increases with load?
c What would you expect the extension to be for each of the following loads?
  i 4 N
  ii 6 N
d As more weights are added, the extension increases. How is the spring affected if it is taken past its elastic limit?

6 A hovercraft has a mass of 1600 kg. It is hovering several centimetres above some water. It is supported by a cushion of air which covers an area of 8 m². The air cushion is kept at a pressure greater than atmospheric pressure. (The difference between the air cushion pressure and atmospheric pressure is called the excess pressure.)

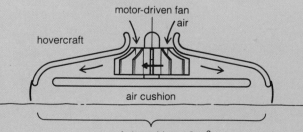

a What job does the motor-driven fan do?
b If the Earth's gravitational field strength is 10 N/kg, what is the weight of the hovercraft in newtons?
c What upward force must the air cushion exert to keep the hovercraft up?

d What is the excess pressure in the air cushion?

7 A model car has a mass of 4 kg. It starts from rest with an acceleration of 3 m/s².

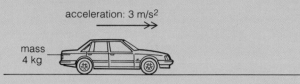

a What is the force on the the car in newtons?
b What is the speed of the car after 5 seconds?
c When the moving car reaches a speed of 15 m/s, it collides with, and sticks to, another identical model car which is stationary:

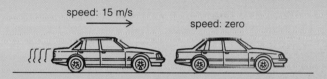

  i What is the momentum of the moving car before the collision?
  ii What is the momentum of the combination after the collision?
  iii What is the speed of the combination after the collision?

8 In the diagram below, someone is swinging a ball round on the end of a piece of string.

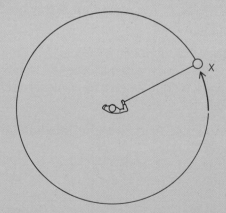

a What name is given to the force needed to make the ball move in a circle?
b Copy and complete the diagram to show where the ball will travel if the string breaks when the ball is at point X.

2.1 Work and energy

Who is doing most work?

In everyday language, 'work' can mean anything from writing an essay to digging the garden.
To scientists and engineers, work has an exact meaning:
Work is done whenever a force moves.

The unit of work is the **joule (J)**.
1 joule of work is done when a force of 1 newton moves a distance of 1 metre (in the direction of the force).

There is an equation for calculating work:

Work done = force × distance moved (in the direction of the force)

For example:

6 J of work is done when a force of 2 N moves 3 m
12 J of work is done when a force of 4 N moves 3 m
24 J of work is done when a force of 4 N moves 6 m
and so on.

Larger units of work are the **kilojoule** and the **megajoule**:

1 kilojoule (kJ) = 1000 J
1 megajoule (MJ) = 1 000 000 J

Work done . . .	
in shutting a door	5 J
in throwing a ball	20 J
in climbing the stairs	1 kJ
in loading a lorry	1 MJ

Energy

Things have energy if they can do work.
A tankful of petrol has energy, so does a stretched spring. Each can be used to make something move. In each case, you can think of the energy as a promise of work to be done in the future. There are several different types of energy.

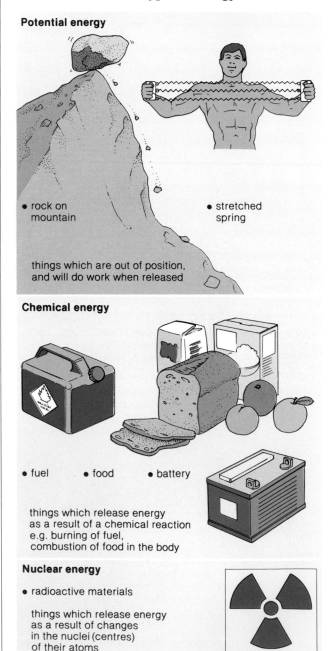

Potential energy

- rock on mountain
- stretched spring

things which are out of position, and will do work when released

Chemical energy

- fuel   - food   - battery

things which release energy as a result of a chemical reaction e.g. burning of fuel, combustion of food in the body

Nuclear energy

- radioactive materials

things which release energy as a result of changes in the nuclei (centres) of their atoms

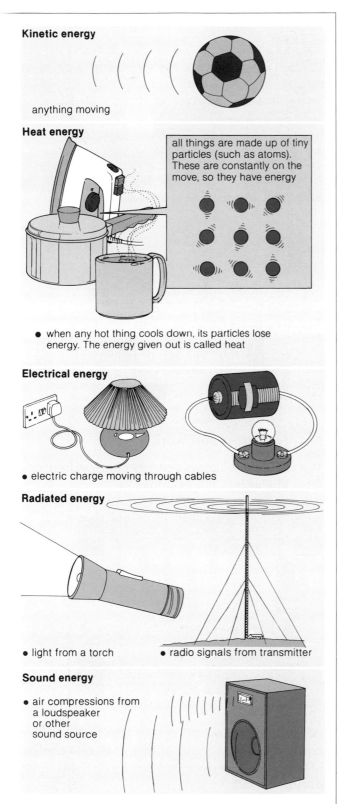

Kinetic energy

anything moving

Heat energy

all things are made up of tiny particles (such as atoms). These are constantly on the move, so they have energy

- when any hot thing cools down, its particles lose energy. The energy given out is called heat

Electrical energy

- electric charge moving through cables

Radiated energy

- light from a torch
- radio signals from transmitter

Sound energy

- air compressions from a loudspeaker or other sound source

How much energy?

The unit of energy is the joule (J).
100 000 joules is the energy you could get from....

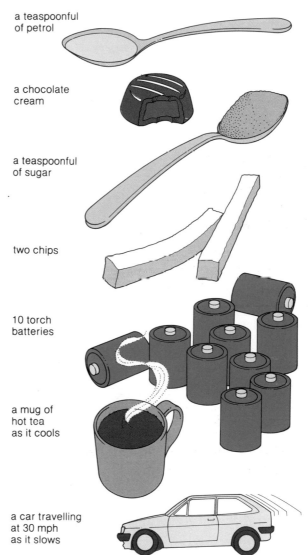

a teaspoonful of petrol

a chocolate cream

a teaspoonful of sugar

two chips

10 torch batteries

a mug of hot tea as it cools

a car travelling at 30 mph as it slows

1 How much work is done when:
a a 6 N force moves 3 m?
b a 12 N force moves 0.5 m?
c a 10 N force moves 10 mm?

2 In the 'How much energy?' chart above, which of the items has:
a CHEMICAL energy?
b POTENTIAL energy?
c KINETIC energy?

3 Which is likely to release most energy:
a burning a can-full of petrol, or dropping it?
b catching a falling apple, or eating it?

2.2 Energy changes

A super-heavyweight and his daily menu:

1 grapefruit	1 bowl of cornflakes
7 pints of milk	12 eggs
8 steaks	1 kg cheese
30 slices of bread	1 kg butter
4 tins of pilchards	2 tins of baked beans
1 rice pudding	1 pot of honey

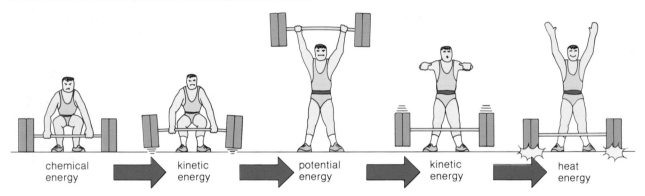

chemical energy → kinetic energy → potential energy → kinetic energy → heat energy

The weightlifter's menu is not a healthy diet for most people but what you might be eating if you were breaking Olympic weightlifting records. Much of the food is used for muscle building. But some provides energy for the lift.

The weightlifter has to supply about 4000 J to lift the weights above his head. The energy is stored in the body as chemical energy. However, during the lift, it is changed into other forms.

When the weights hit the ground, the impact makes the particles move faster. This means that ground, air and weights are all a little warmer than before. The 4000 J of energy has become heat energy.

From the first stage to the last, the *type* of energy changes, but the total *amount* of energy stays the same.

This is an example of the **law of conservation of energy**:

Energy can change from one type to another, but it cannot be made or destroyed.

Energy change and work

Work is done whenever energy changes from one form to another. For example:

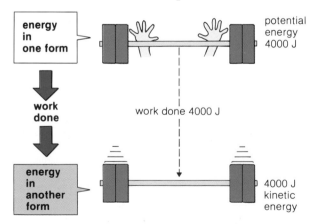

energy in one form → work done → energy in another form

potential energy 4000 J

work done 4000 J

4000 J kinetic energy

After the weights are dropped, 4000 J of potential energy is changed into 4000 J of kinetic energy: 4000 J of work is done in speeding up the weights during the fall:

work done = energy changed

Energy converters

An energy converter changes energy from one type into another.

The weightlifter is an energy converter. So are you.

So is each of the following:

an iron changes **electrical** energy into **heat** energy

brakes change **kinetic** energy into **heat** energy

telephones change **sound** energy into **electrical** energy

. . . into **sound** energy

a swing changes **kinetic** energy into **potential** energy into **kinetic** energy into **potential** energy into . . .

Most energy converters lose some of their energy as heat.

For example, to do 4000 J of work, the weightlifter has to release about 25 000 J of stored energy by 'burning' food he has eaten. The spare 21 000 J makes him hot – which is why he sweats.

When things move, they often lose energy because of air resistance of friction. The swing slowly loses its energy as heat.

As it pushes through the air, it slows down and the air particles (molecules) speed up.

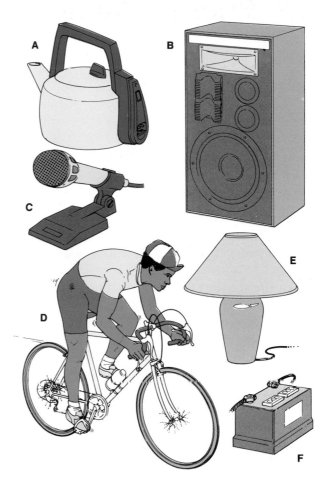

1 Which of the above:
a changes CHEMICAL energy into KINETIC energy?
b changes ELECTRICAL energy into HEAT energy?
c changes CHEMICAL energy into ELECTRICAL energy?
d changes ELECTRICAL energy into RADIATED energy?
e changes ELECTRICAL energy into SOUND energy?

2 A pole-vaulter has 3500 J of energy when he crosses a bar 7 metres high.
How much kinetic energy will he have just before he reaches the ground?

3 This is an unusual way of describing the first half hour of someone's day:

   gets up at 7:00 a.m.
   gains chemical energy
   gains heat energy
   leaves house
   kinetic energy rises
   jumps on vehicle
   potential energy rises
   kinetic energy rises
   kinetic energy fall to zero
   potential energy falls to zero

Rewrite, to show what could actually be happening to the person.

2.3 Potential and kinetic energy

Moving fast, high above the Earth, travelling at around 8 kilometres per second, the space shuttle has a great deal of both potential and kinetic energy.

When the shuttle re-enters the atmosphere, that energy is changed into heat energy. But how much energy?

If you were designing the thousands of heat-resistant tiles that protect the surface of the shuttle, you would need to know.

Potential and kinetic energies can be calculated, but it is easier to start with something lighter and slower than the shuttle.

Like a stone lifted above the ground, or thrown.

The same by any route

When calculating potential energy, it is the height lifted against gravity that matters, not the actual distance moved.

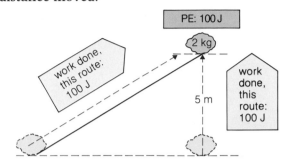

When the stone is lifted above the ground it gets the same potential energy whether it is lifted straight up or pulled up the slope. It takes less force to pull the stone up the slope. But the distance is further.

Result: the work done is the same by either route.

Calculating potential energy (PE)

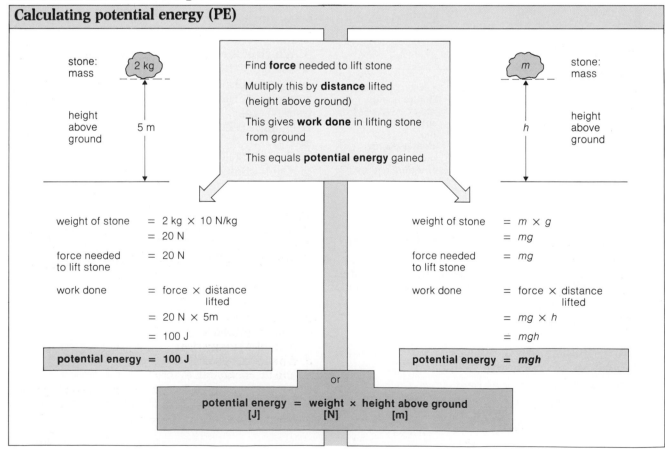

stone: mass	2 kg
height above ground	5 m

Find **force** needed to lift stone

Multiply this by **distance** lifted (height above ground)

This gives **work done** in lifting stone from ground

This equals **potential energy** gained

stone: mass	m
height above ground	h

weight of stone	=	2 kg × 10 N/kg
	=	20 N
force needed to lift stone	=	20 N
work done	=	force × distance lifted
	=	20 N × 5m
	=	100 J

potential energy = 100 J

weight of stone	=	$m \times g$
	=	mg
force needed to lift stone	=	mg
work done	=	force × distance lifted
	=	$mg \times h$
	=	mgh

potential energy = mgh

or

potential energy = weight × height above ground
[J]   [N]   [m]

Calculating kinetic energy (KE)

To calculate the kinetic energy of a stone, mass m and velocity v, use the equation:

$$\text{kinetic energy} = \tfrac{1}{2}mv^2$$

Look at the box if you want to find out why.

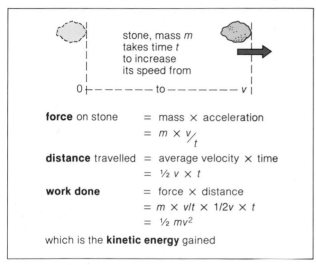

stone, mass m takes time t to increase its speed from 0 to v

force on stone	= mass × acceleration
	= $m \times v/t$
distance travelled	= average velocity × time
	= $\tfrac{1}{2}v \times t$
work done	= force × distance
	= $m \times v/t \times 1/2v \times t$
	= $\tfrac{1}{2}mv^2$

which is the **kinetic energy** gained

For example, take a stone of mass 2 kg travelling at a velocity of 6 m/s:

$$\text{kinetic energy} = \tfrac{1}{2} \times 2 \times 6^2 \ \text{J} = 36\,\text{J}$$

Adding energies

What happens if you lift the 2 kg stone 5 m above the ground, *and* throw it at 6 m/s?
The stone has 100 J of potential energy and 36 J of kinetic energy
This gives it a total of 136 J. The energies add together easily because energy is not a vector and you don't have to allow for direction.

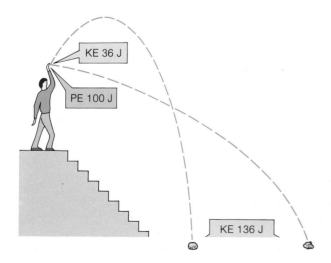

KE 36 J

PE 100 J

KE 136 J

Just before the stone hits the ground, all this energy will be kinetic energy (assuming the stone doesn't lose energy because of air resistance). The stone ends up with the same kinetic energy – and the same speed – whether it is thrown upwards, downwards, or sideways. The direction of throw makes no difference.

Assume $g = 10\,\text{N/kg}$.

1

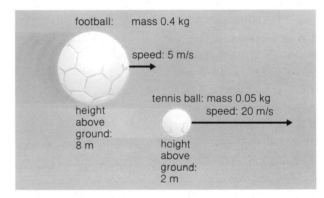

football: mass 0.4 kg
speed: 5 m/s
height above ground: 8 m

tennis ball: mass 0.05 kg
speed: 20 m/s
height above ground: 2 m

Use the information in the diagram to work out:
a the potential energy of each ball;
b the kinetic energy of each ball;
c the kinetic energy of each ball just before it reaches the ground.

2 Here is some information about a space shuttle in orbit:

mass:	100 tonnes
speed:	8 km/s
height above ground:	100 km

a What is the mass in kg?
b What is the speed in m/s?
c What is the height above the ground in metres?
d What is the potential energy of the shuttle?
e What is its kinetic energy?
f How much energy must it lose before coming to a halt on the ground?
g What happens to this energy?

2.4 Engines

For smooth running, there's nothing quite like a steam engine under the bonnet. And this 1920 Stanley Steamer was the last production car to have one. You lit the fuel, waited 20 minutes for the water to boil, and were off – at anything up to 80 mph.

The job of all engines is to make things move. They do so by turning energy in their fuel into kinetic energy. Steam is still used in the world's largest engines. But for modern road vehicles, there are other possibilities:

Petrol engines

Petrol engines use the force of an exploding petrol/air mixture to produce motion. Most engines follow a cycle of four up and down strokes. During one stroke, the mixture is exploded, and the force moves a piston in a cylinder. During the other three strokes, burnt gases are removed and a fresh mixture prepared for explosion.

Diesel engines

Diesel engines also follow a four-stroke cycle. But they use fuel oil instead of petrol.
Air entering the engine is compressed so much that it becomes very hot. Fuel oil is squirted straight into the cylinders. The oil ignites as soon as it meets the hot air. So diesel engines don't need spark plugs.

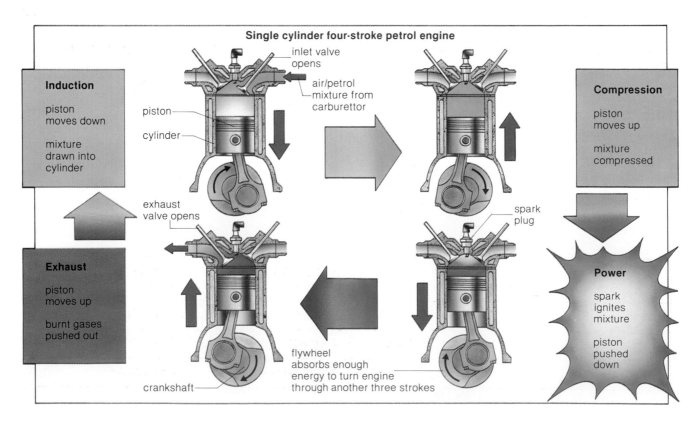

Single cylinder four-stroke petrol engine

Induction

piston moves down

mixture drawn into cylinder

inlet valve opens

air/petrol mixture from carburettor

piston

cylinder

Compression

piston moves up

mixture compressed

spark plug

Power

spark ignites mixture

piston pushed down

Exhaust

piston moves up

burnt gases pushed out

exhaust valve opens

flywheel absorbs enough energy to turn engine through another three strokes

crankshaft

How engines compare

The fuel for petrol and diesel engines comes from crude oil (and Earth's resources are slowly running out). Electric motors don't use fuel directly. But they need an electric current, and that may well come from a fuel-burning power-station. However, there is one type of engine that doesn't have these problems....

	petrol	diesel	electric	human
engine				
energy source	petrol	fuel oil	battery	food
advantages	• good performance for low cost and weight	• uses less fuel than petrol engine	• quiet • no air pollution from motor • works well at low speeds	• quiet • little air pollution • works well at low speeds • fuel can be regrown
disadvantages	• work poorly at low speeds, need clutch and gearbox • pollute atmosphere, fuels cannot be replaced	• heavier, more expensive than petrol engine • work poorly at low speeds, need clutch and gearbox • pollute atmosphere, fuels cannot be replaced	• batteries very heavy and can't store much energy	• poor at moving heavy loads quickly

1 Vans with diesel engines cost more to buy than vans with petrol engines. But they use less fuel. Electric-powered vans are slower than either, and very expensive. The table below gives some information about the cost of hiring each type of van.
a Which van has the lowest fuel or energy bill for a 100 kilometre journey?
b How long does each van take to travel 100 km?
c What is the total cost of a 100 km journey in each van? Remember to include hire charges as well as fuel or energy costs.
d What is the total cost of a 300 km journey made in each van?

e If you wanted to use as little energy as possible, which van would you select?
f If you were interested in keeping your costs down, which van would you select for a short journey? Which would you select for a long journey?

2 Petrol and diesel engines take in fuel and produce movement.
a What type of energy do they take in?
b What happens to this energy inside the engine?
c What type of energy do they give out?

	Hire charge van and driver	Energy cost per 100 km	Average speed
Van A: petrol	£20 (+£5 per hour)	£6 (fuel)	50 km/hour
Van B: diesel	£24 (+£5 per hour)	£4 (fuel)	50 km/hour
Van C: electric	£30 (+£5 per hour)	£2 (recharging)	25 km/hour

2.5 Efficiency and power

Like most other engines, the human engine is a wasteful user of fuel.

If you are pedalling hard on a bike, for every 100 joules of energy released from your food, only about 15 joules is used to work. The rest is turned into heat – you sweat to get rid of it.

So the efficiency of a cyclist $= \dfrac{15}{100}$.

This can be written as a percentage: 15%.
Here's how other engines compare:

For every 100 J of energy put into a . . .		the work done is . . .	the efficiency is . . .
petrol engine		25 J	25%
diesel engine		35 J	35%
electric motor		80 J	80%
human engine		15 J	15%

Only the electric motor seems to have a high efficiency. But that value is deceiving. Electricity for the motor has to be generated, and the efficiency of that process is only about 30%.

Why are engines such poor energy converters? It isn't the fault of the manufacturers. They constantly seek new ways of reducing engine friction and improving fuel burning. The problem lies with the laws which govern how moving particles (such as atoms and molecules) behave. For more on this, see page 71.

Power

A small car engine can do just as much work as a Land Rover engine. But it takes longer to do it. The Land Rover engine has more power than the small car engine; it can do more joules of work *every second*.

Power is measured in **watts (W)**.
An engine with a power output of 1000 watts can do 1000 joules of work every second.
An engine with a power output of 2000 watts can do 2000 joules of work every second. And so on.

Power is calculated using the equation:

$$\textbf{power} = \frac{\textbf{work done}}{\textbf{time taken}} \quad \textbf{or} \quad \frac{\textbf{energy changed}}{\textbf{time taken}}$$

For example:
800 J work done ÷ 2 s time taken = 400 W power output

Powers are sometimes given in **kilowatts** or **megawatts**:
1 kilowatt (kW) = 1000 watts
1 megawatt (MW) = 1 000 000 watts

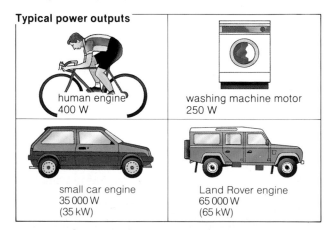

Typical power outputs

human engine 400 W

washing machine motor 250 W

small car engine 35 000 W (35 kW)

Land Rover engine 65 000 W (65 kW)

Power values can be used to calculate efficiency:

$$\textbf{efficiency} = \frac{\textbf{power output}}{\textbf{power input}}$$

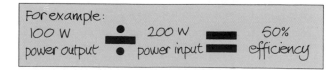

For example:
100 W power output ÷ 200 W power input = 50% efficiency

How to measure your power output:

Assume $g = 10\,\text{N/kg}$.

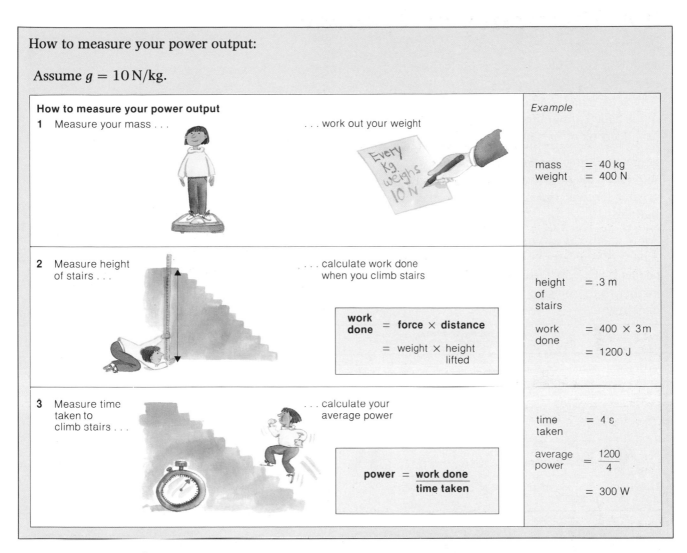

How to measure your power output		Example
1 Measure your mass . . . . . . work out your weight		mass = 40 kg weight = 400 N
2 Measure height of stairs . . . . . . calculate work done when you climb stairs **work done = force × distance** = weight × height lifted		height of stairs = .3 m work done = 400 × 3 m = 1200 J
3 Measure time taken to climb stairs . . . . . . calculate your average power **power = $\dfrac{\text{work done}}{\text{time taken}}$**		time taken = 4 s average power = $\dfrac{1200}{4}$ = 300 W

Assume $g = 10\,\text{N/kg}$.

1 The cheetah is the fastest creature on land. A typical cheetah, at full speed, has a power output of 1000 W, and an efficiency of 15%.

Mike is possibly the slowest creature on land. When he works in the garden (which isn't very often), his power output is 100 W, and his efficiency 5%.
Calculate:
a the work done by the cheetah in 1 second;
b how long it takes Mike to do the same amount of work.
In unfolding his garden chair, Mike gets 2000 J of energy from the food he has eaten.
Calculate:
c how much work he does.
Write down what happens to the rest of the energy released.
2 A skier has a mass of 50 kg.
It takes her 40 seconds to climb 20 m (vertically) to the top of a slope.
Calculate:
a her weight;
b the work done when she climbs the slope;
c her average power output.
Why, in reality, will she have to do more work than you have calculated?

2.6 Machines

Anything which makes forces more convenient to use is called a **machine**. It may be as complicated as a gearbox or as simple as a pair of scissors.

Some machines are **force magnifiers**.
A pair of pliers for example. These give you a greater force at the jaws than you put in at the handles.

Some machines are **movement magnifiers**.
A bicycle for example. One downward push on the pedals takes you forward over 3 metres – much further than one step would take you if you were walking.

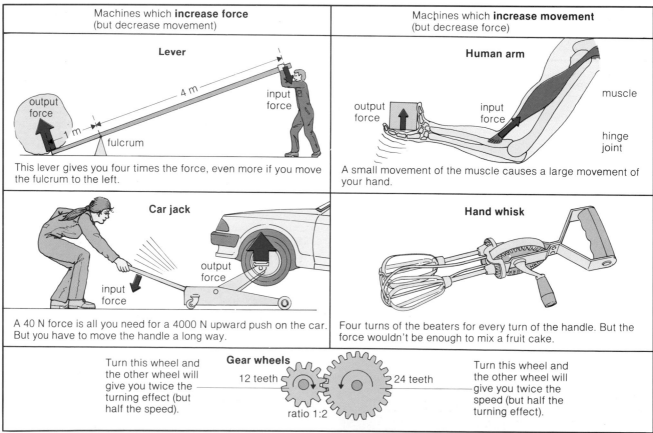

Machines which **increase force** (but decrease movement)	Machines which **increase movement** (but decrease force)
Lever — 4 m, output force, input force, fulcrum, 1 m — This lever gives you four times the force, even more if you move the fulcrum to the left.	**Human arm** — output force, input force, muscle, hinge joint — A small movement of the muscle causes a large movement of your hand.
Car jack — output force, input force — A 40 N force is all you need for a 4000 N upward push on the car. But you have to move the handle a long way.	**Hand whisk** — Four turns of the beaters for every turn of the handle. But the force wouldn't be enough to mix a fruit cake.

Turn this wheel and the other wheel will give you twice the turning effect (but half the speed).

Gear wheels — 12 teeth — 24 teeth — ratio 1:2

Turn this wheel and the other wheel will give you twice the speed (but half the turning effect).

No machine magnifies both force *and* movement. If it did, you would get more work out of the machine than you put into it:

$$\text{work} = \text{force} \times \text{distance moved}$$

So, if a machine *increases* force, it must *decrease* distance moved. And vice versa.

Most machines have friction between their moving parts. This means that some energy is lost as heat. And you get less useable energy out of the machine than you put in.
In other words:
the work output is less than the work input; the efficiency is less than 100%.

Measuring the efficiency of a machine

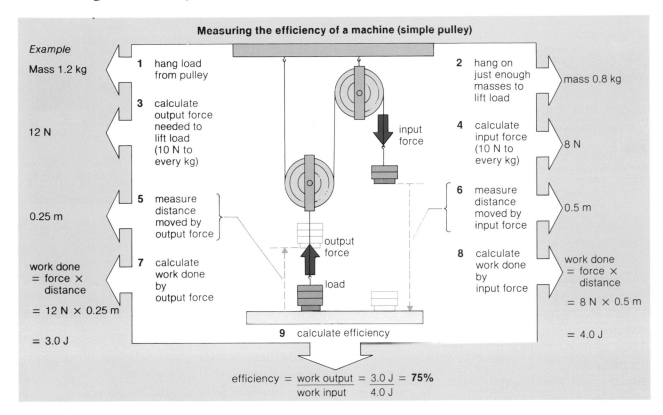

Measuring the efficiency of a machine (simple pulley)

Example

Mass 1.2 kg

1 hang load from pulley

2 hang on just enough masses to lift load — mass 0.8 kg

12 N

3 calculate output force needed to lift load (10 N to every kg)

4 calculate input force (10 N to every kg) — 8 N

0.25 m

5 measure distance moved by output force

6 measure distance moved by input force — 0.5 m

work done = force × distance = 12 N × 0.25 m = 3.0 J

7 calculate work done by output force

8 calculate work done by input force

work done = force × distance = 8 N × 0.5 m = 4.0 J

9 calculate efficiency

$$\text{efficiency} = \frac{\text{work output}}{\text{work input}} = \frac{3.0 \text{ J}}{4.0 \text{ J}} = \mathbf{75\%}$$

---

1 Make two lists to show which of these are FORCE magnifiers and which are MOVEMENT magnifiers:

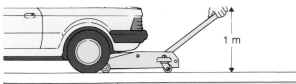

Nail clippers   Kitchen scales
Bottle opener   Hole punch
Pincers   Hand drill
Door handle   Tin opener

2 Dave and Sue are about to use a car jack to lift their car. Here is some information about the car and the jack:

Car	Car jack
force needed to lift rear of car = 2000 N	distance moved by handle during one stroke = 1 m
	input force to lift rear end of car = 20 N

How much work must be done to lift the rear of the car 1 metre?
Sue says that it will take at least 100 strokes to lift the rear of the car 1 metre. Dave says it won't. Who is right? And why?

3 A mechanic uses a pulley to lift an engine out of a car.
The engine weighs 500 N, and is lifted 2 m.
The mechanic uses a force of 200 N on the rope, which is pulled downwards a distance of 10 m.
Calculate:
a the work done on the engine in lifting it;
b the work done by the mechanic;
c the efficiency of the pulley.

4 A, B, and C are three gear wheels.

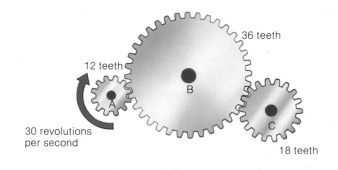

36 teeth

12 teeth

30 revolutions per second

18 teeth

A is rotated 30 times per second.
Sketch the diagram.
Show which direction each wheel is turning.
Write down the number of times per second each wheel is turning.

65

2.7 Liquid machines

In the digger in the photograph, the shovels are moved by high-pressure oil in flexible pipes.

The basic idea is simple:

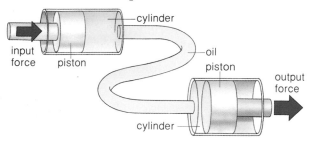

A pump or piston pushes on the oil at one end. The oil moves a piston at the other end.
The idea works because liquids have two special features:
1 They cannot be squashed – they are virtually incompressible.
2 If a trapped liquid is put under pressure, the pressure is transmitted to all parts of the liquid.

Machines which use trapped liquids in this way are called **hydraulic** machines.

Force magnifiers

Most hydraulic machines are force magnifiers. They give out more force than is put in. This happens because the output piston is larger than the input piston. To find out why, study the simple hydraulic jack on the opposite page.

Like all other machines, hydraulic machines can't give out more *work* than is put in:

work = force × distance moved

So, if a machine *increases* force, it *decreases* the distance moved. In other words, the output force doesn't move as far as the input force.

Hydraulics in action

Caterpillar tracks: driven by hydraulic motors fed with high-pressure oil from a pump.

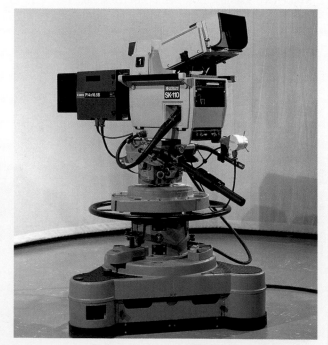

TV camera: hydraulic action raises and lowers the camera on its mount.

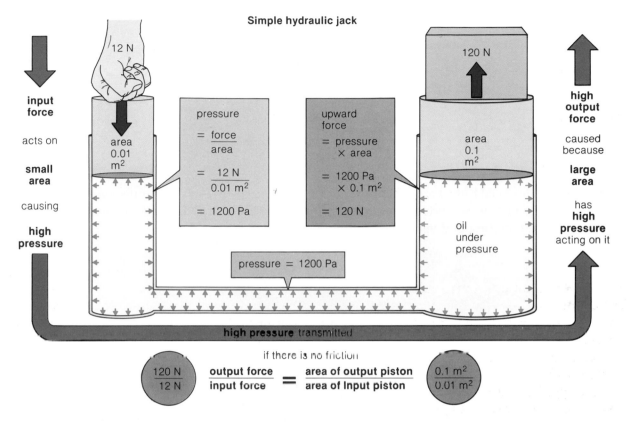

Simple hydraulic jack

input force acts on small area causing high pressure

12 N

area 0.01 m²

pressure
= force / area
= 12 N / 0.01 m²
= 1200 Pa

upward force
= pressure × area
= 1200 Pa × 0.1 m²
= 120 N

pressure = 1200 Pa

120 N

area 0.1 m²

oil under pressure

high output force caused because large area has high pressure acting on it

high pressure transmitted

if there is no friction

120 N / 12 N

output force / input force = area of output piston / area of input piston

0.1 m² / 0.01 m²

---

1 Decide whether each of the following statements is TRUE or FALSE.

a Liquids are virtually impossible to compress.

b If a trapped liquid is under pressure, the pressure is transmitted throughout the liquid.

c A machine can give out more force than is put into it.

d A machine can give out more work than is put into it.

2 The diagram shows the basic layout of a car's hydraulic brake system.

The brake pedal is a simple lever which magnifies the force from the foot.

The disc is attached to a road wheel. It slows down when the brake pad is pushed against it.

(In an actual car, the disc would be squeezed between two pads. And the fluid would be piped to brakes on all four wheels.)

Using the information in the diagram, calculate:

a the force on the piston A;

b the pressure on the fluid at B;

c the pressure on the fluid at C;

d the force on the brake pad.

Say whether the force on the pad would be MORE or LESS:

e if the connecting point O was nearer the pivot;

f if the area of piston A was less.

Can you think of any reason why water isn't used in the brake system?

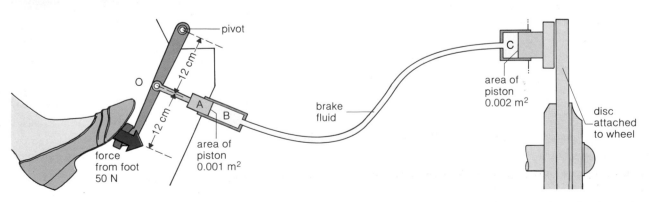

pivot

12 cm

O

12 cm

force from foot 50 N

A

B

area of piston 0.001 m²

brake fluid

C

area of piston 0.002 m²

disc attached to wheel

2.8 Energy resources

How energy is used in the UK

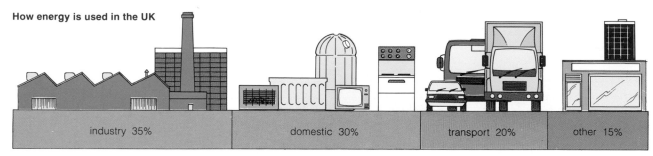

industry 35%	domestic 30%	transport 20%	other 15%

Industrial societies need huge amounts of energy. Most comes from fuels which are burnt in power-stations, factories, homes, and vehicles.

The energy in fuels originally came from the Sun. So did the energy in the foods we eat.

...in fuels

Energy from the Sun...

...in foods

natural gas

coal

oil

plants take in radiant energy from the Sun as they grow

fossil fuels formed from decayed plants and animals which lived over 200 million years ago

animals eat plants

When food is eaten, more can be grown. It is a **renewable** energy source. But coal, oil and natural gas can't be renewed. And the Earth's supply of them is gradually running out. At present rates of use, there is probably:
– enough oil and natural gas left to last for 50–100 years;
– enough coal left to last for 200–300 years.

Oil is an especially useful material:
– it contains fuels like petrol and diesel. These aren't very bulky, so they can be carried and used in vehicles.
– it is the raw material from which most plastics are made.
With oil supplies running out, it makes sense to save what is left for transport and plastics.

Alternative energy sources

Here are some alternatives to coal, oil and natural gas. Most are renewable. Once used, the energy is replaced naturally.

Wind energy Giant wind turbines (windmills) turn electrical generators. *Aerogenerators* like this can be positioned in a large group called a *wind farm*.	*For:* renewable energy source. *Against:* aerogenerators large, costly and noisy, with relatively low power output. Not enough wind in many areas.
Hydroelectric energy Rivers fill a lake behind a dam. Fast-flowing water from the lake turns generators.	*For:* renewable energy source *Against:* few areas of the world suitable.
Tidal energy A dam is built across an estuary. A lake behind the dam fills up at high tide, and empties at low tide. Fast-flowing water turns generators. The Earth's movement is the source of tidal energy. The Moon's gravitational pull causes 'bulges' of sea water on the Earth's surface. As the Earth rotates, each part passes in and out of a bulge – the tide rises and falls.	*For:* renewable energy source. *Against:* very expensive to set up; few areas suitable. 
Solar energy Mirrors and panels are used to capture the Sun's radiant energy – usually as heat.	*For:* renewable energy source. *Against:* continuous sunshine needed.
Nuclear energy Radioactive materials naturally release heat. A nuclear reactor speeds up the process. The heat is used to generate electricity (see page 209).	*For:* small amounts of nuclear fuel give large amounts of energy. *Against:* nuclear radiation extremely dangerous. High safety standards needed. Waste materials from power-stations stay radioactive for thousands of years.
Geothermal energy Water is heated by the hot rocks which lie many miles beneath the Earth's surface. The heat in the rocks comes from radio-active materials naturally present in the Earth.	*For:* renewable energy source. Huge quantities of energy available. *Against:* deep drilling very difficult and expensive.
Biomass energy Fast-growing plants, or *biomass*, used to make alcohol. Alcohol used as a fuel, like petrol.	*For:* renewable energy source. *Against:* huge land areas needed to grow plants; this may upset the balance of nature.

1 What is the difference between a renewable and a non-renewable source of energy?

2 COAL, OIL, BIOMASS, WIND, TIDES, NATURAL GAS Which of these are renewable energy sources, and which are not?

3 When you eat a cheese sandwich, you take in energy which came from the Sun. How does the energy pass from:
a the Sun to the bread?
b the Sun to the cheese?

2.9 Energy issues

Disturbing the atmosphere

Fuel-burning power-stations give off huge amounts of carbon dioxide gas. Although plants take in carbon dioxide as part of their life process, more carbon dioxide is being added to the atmosphere than is being removed.

Carbon dioxide acts like the glass in a greenhouse. It traps the Sun's heat. As a result, the Earth is slowly warming up. Scientists call this **global warming** or the **greenhouse effect**. The warming is only slight, but it may have a major effect on world climates (see page 105).

Burning fuels can upset the atmosphere in other ways as well. For example, coal-burning power-stations emit sulphur dioxide gas. This causes **acid rain** which can kill trees and water life, and damage stonework. One solution is to burn expensive low-sulphur coal. Another is to fit costly **flue gas desulphurization (FGD)** units to the power-stations.

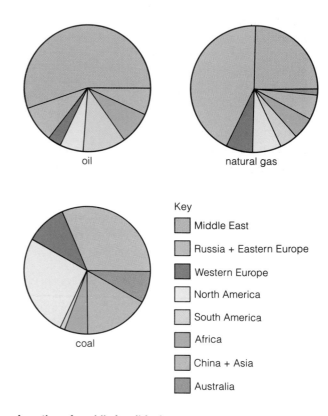

oil

natural gas

Key
- Middle East
- Russia + Eastern Europe
- Western Europe
- North America
- South America
- Africa
- China + Asia
- Australia

coal

Location of world's fossil fuel reserves

Power-stations compared

Here is some data about different power-stations:

Power-station ▶ (1 MW = 1000 kW = 1 000 000 W)	A Coal (non-FGD)	B Combined cycle gas	C Nuclear	D Wind farm	E Large tidal scheme
Power output in MW	1800	600	1200	20	6000
Efficiency (fuel energy → electrical energy)	35%	45%	25%	—	—
The following are on a scale 0–5					
Build cost per MW output	2	1	5	3	4
Fuel cost per kW h output	5	4	2	0	0
Atmospheric pollution per kW h output	5	3	<1	0	0

In a **combined cycle gas turbine power-station**, natural gas is used as the fuel for a jet engine. The shaft of the engine turns one generator. Heat from the jet exhaust is used to make steam to drive another generator.

Nuclear power-stations have an additional cost: the cost of decommissioning them (closing them down and dismantling them at the end of their working life). This can be almost as much as building them in the first place.

Energy spreading

In a typical fuel-burning or nuclear power-station, the generators are turned by huge turbines (fans), driven by steam from a boiler (see page 192). Really, the boiler and turbines are a giant engine. The fuel's energy is released as heat, and this is used to produce motion.

Like other engines, the engines in power-stations waste more energy than they deliver. The energy is lost as heat in the cooling water and waste gases. For example, the efficiency of a typical coal-burning power-station is only about 35%. This is what happens to the energy:

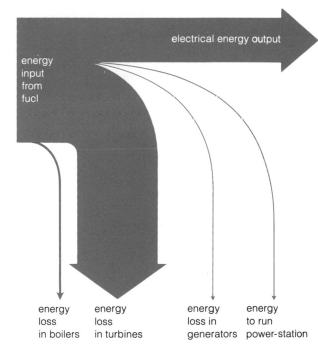

energy input from fuel

electrical energy output

energy loss in boilers

energy loss in turbines

energy loss in generators

energy to run power-station

Engineers try to make power-stations as efficient as possible. But once energy is in the form of heat, it is impossible to use all of it to produce motion:

Heat is the energy of randomly moving particles (such as atoms and molecules). It has a natural tendency to spread out. As heat spreads, it becomes more and more difficult to turn it into other forms of energy. In other words, the energy becomes less and less useful. For example, the energy from hot, burning fuel can be used to make steam which will drive a turbine. But if the heat is spread out so that it warms a huge tankful of water, it becomes much more difficult to use the heat from the warm water to drive the turbine.

1 The pie-charts on the opposite page compare fossil fuels in different regions of the world:
a Which region has the largest oil reserves?
b What other fossil fuel does this region have plentiful reserves of?
c The USA needs to import oil but not coal. Can you suggest reasons why?
d Which regions of the world do you think consume most fossil fuels?
e Why would it be desirable to reduce the world's consumption of fossil fuels?

2 Look at the table of data on the opposite page about five different power-stations A–E:
a Which power-station has the greatest efficiency?
b The efficiency of the nuclear power-station is 25%. Explain what this means.
c Which power-station has the highest fuel cost (per kW h output)?
d Which power-station cost most to build?
e Why do two of the power-stations have a zero rating for fuel costs and atmospheric pollution?
f Which power-station produces most atmospheric pollution per hour? What can be done to reduce this problem?

3 The table below gives data about the power input and power losses in two power-stations, X and Y:
a Where is most energy wasted?
b In what form is this wasted energy lost?
c What is the electrical power output of each station? (You can assume that the table shows all the power losses in each station.)
d What equation is used to calculate efficiency? (You can find this on page 62.)
e Calculate the efficiency of each power-station.

Power-station	X Coal	Y Nuclear
Power input from fuel in MW	5600	5600
Power losses in MW: —in reactors/boilers —in turbines —in generators Power to run station in MW	600 2900 40 60	200 3800 40 60
Electrical power output in MW	?	?

The Sun is powered by nuclear fusion (see page 243). One day, fusion reactors may be built on Earth. Their hydrogen fuel is plentiful, and they should produce little radioactive waste. But developing them will be slow and expensive.

2.10 Saving energy

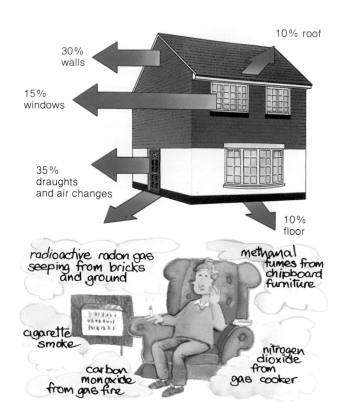

Escaping heat

Lost heat can cost a family well over £500 a year in fuel bills. The illustration on the right shows how the heat escapes from a house.

New air for old

Stopping draughts and air changes saves most on the fuel bills. But it can put your health – and even your life – at risk. If rooms are tightly sealed, the oxygen used by people and fires isn't replaced. And dangerous chemicals collect in the air.

For safety's sake, there should be at least one complete change of air in a room every hour. In a draughty house, there may be 15 more. This is good for clearing the air, but it makes the house very expensive to heat.

Low-energy families

The Birkebaeks live in Denmark. They had this low-energy house specially designed and built for them. In winter, the power of a one-bar electric heater is enough to heat the whole house. The house cost them over £250 000, including the land and architect's fees.

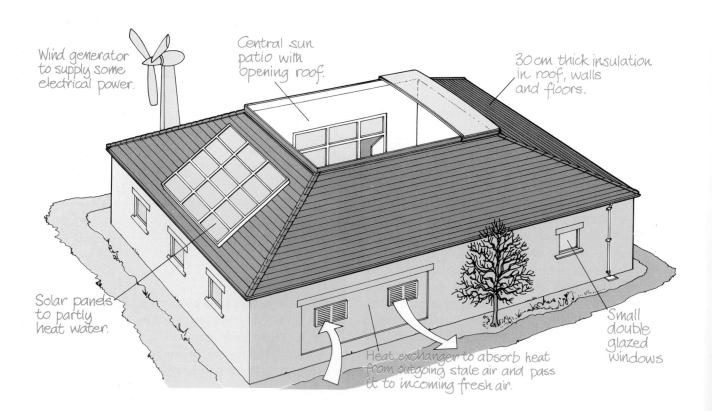

The Cottrells live in Newcastle. Their house is sandwiched between two others. This helps stop some heat loss. But the house is still cold and draughty. They would like to replace the doors and windows with modern ones that fit properly. But they can't afford to. They save energy by heating only one room. It means that their bedroom is very cold in winter, but they can pay the fuel bills.

When power-stations generate electricity, much of the energy from their fuels is lost as heat. The heat warms up the cooling water that flows through the power-station. Usually, the energy is

wasted. But one idea is to use the warmed water to heat buildings nearby.

But there is a problem. Power-station water isn't quite hot enough for room heating. To produce hotter water, each power-station would have to lose some of its power output. And that would make it more difficult for the electricity companies to stay in profit.

Going to waste

What causes the greatest loss of heat in most houses?
Why can it be dangerous to stop this loss completely?

Design a laboratory experiment to find out which loses most heat – a terraced house, a semi-detached house, or a detached house.

How would you set up the experiment and what measurements would you make?

When comparing houses, how would you make sure that your results were 'fair'?

How would you modify your experiment to test whether:
● big houses lose more heat than small houses;
● houses with big windows lose more heat than houses with small windows?

terraced

semi-detached

detached

2.11 Moving particles

Like the land, sea and air in the photograph, everything around you is either a solid, a liquid or a gas. The **kinetic theory of matter** tries to explain how solids, liquids and gases behave. Read the evidence, then judge for yourself.

The kinetic theory

According to this theory, solids, liquids, and gases are made up of tiny particles (such as atoms and molecules). These are far too small to see. They are constantly on the move. When close, they attract each other strongly, and may even stick together.

A **solid**, such as rock, has a fixed shape and volume. Its particles are held close together by strong forces of attraction. They vibrate to and fro, but can't change positions.

A **liquid**, such as water, has a fixed volume, but can flow to fill any shape. Its particles are still close together. But they vibrate so vigorously that the forces of attraction can't hold them in fixed positions.

A **gas**, such as air, has no fixed shape or volume. It quickly fills any space available. Its particles move about at high speed – colliding with each other and the walls of their container. They attract each other hardly at all, and are too spread out to stick together.

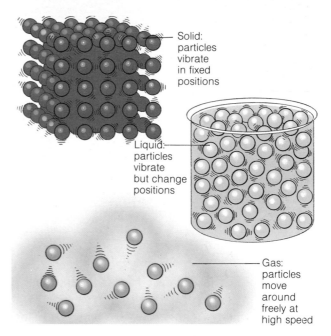

Solid: particles vibrate in fixed positions

Liquid: particles vibrate but change positions

Gas: particles move around freely at high speed

Moving particles – the evidence

Look at smoke through a microscope, and you see a very interesting effect. Glinting in the light, thousands of tiny 'bits' of smoke wobble about in zigzag paths as they drift through the air. This is called Brownian motion, after the scientist Robert Brown, who first noticed it.

The kinetic theory explains the effect as follows: The smoke particles are just big enough to be seen, but small and light enough to be bumped and jostled by the invisible gas molecules in the air around them. They make their jerky movements because gas molecules keep crashing into them.

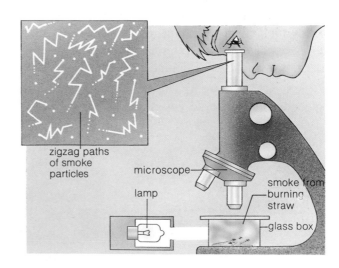

zigzag paths of smoke particles

microscope

lamp

smoke from burning straw

glass box

Wandering particles

Sit and watch the sugary bits on top of a trifle for a day or two, and you may notice that the colours run. According to the kinetic theory, loose particles from the colouring material slowly wander through the surrounding material as they are bumped and jostled by its particles. The process is called **diffusion**.

Smells spread by diffusion. Smells are wandering gas molecules which come from whatever it is that happens to be smelling.

Particles – how big?

The diagram shows an experiment you can carry out to estimate the size of a particle of olive oil. The idea is to place a small drop of the oil on to the surface of some water, so that it spreads to form a thin circular layer. When the oil has spread as far as it can, it will be just one particle thick.

First, sprinkle some lycopodium powder over the surface of some clean water in a tray. The powder makes it easier to see the edge of the oil layer.

Next, pick up a small drop of olive oil on the end of a wire. Use a millimetre scale to measure the diameter of the drop, so that you can calculate its volume.

Then, put the oil drop on to the water surface, and see how far it spreads. Measure the diameter of the circular layer and calculate its area.

The volume of the oil drop is exactly the same as the volume of the circular layer, so:

volume of drop = area of layer × thickness

rearranging

$$\frac{\text{thickness}}{\text{(size of particle)}} = \frac{\text{volume of drop}}{\text{area of layer}}$$

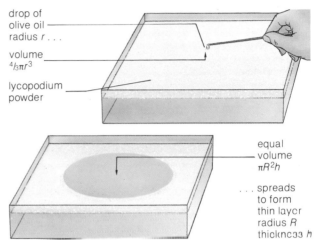

drop of olive oil radius r . . .

volume $\frac{4}{3}\pi r^3$

lycopodium powder

equal volume $\pi R^2 h$

. . . spreads to form thin layer radius R thickness h

Experiments show that particles vary enormously in size. Particles of proteins from living tissue, for example, are many thousands of times larger than the smallest particles. In the case of a medium-sized particle like olive oil, there are around a million to each millimetre. You can use the readings in question 4 to estimate the size for yourself.

1 Say whether each of the following describes a SOLID, a LIQUID or a GAS:
a particles moving about at high speed;
b particles vibrate but can't change positions;
c fixed shape and volume;
d particles vibrate but can change positions;
e no fixed shape or volume;
f fixed volume, but no fixed shape;
g forces of attraction very weak.
2 What do you see if you look at smoke through a microscope?
How does the kinetic theory explain the behaviour of the smoke?
3 Which of the following are examples of *diffusion*:
a water flowing and changing shape;
b ink spreading when dropped in water;
c a smell travelling across a room;
d the particles in a solid vibrating.
4 A drop of olive oil has a volume of 0.3 mm³.
Placed on water, it spreads to form a circular layer of area 200 000 mm². What is the size of a particle of olive oil?

2.12 Temperature

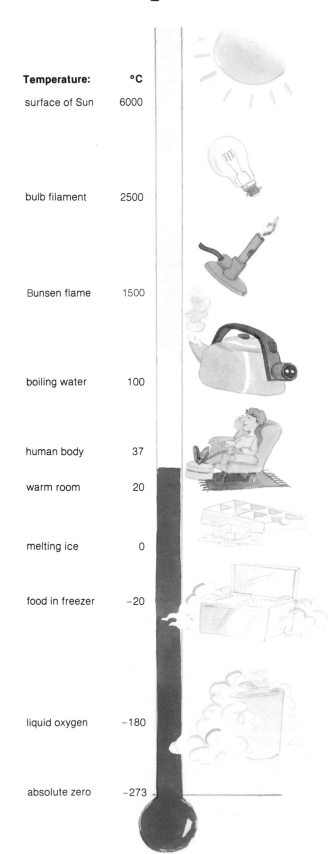

Temperature:	°C
surface of Sun	6000
bulb filament	2500
Bunsen flame	1500
boiling water	100
human body	37
warm room	20
melting ice	0
food in freezer	−20
liquid oxygen	−180
absolute zero	−273

Death Valley, California. The hottest place on Earth. Certainly hotter than ice. But not as hot as boiling water. You can tell that from its **temperature**.

The Celsius scale

The **Celsius** or centigrade scale is the most widely used temperature scale.

On this scale:
The numbers are called **degrees Celsius**, and written °C.
Pure ice melts at 0 °C.
Pure water boils at 100 °C.
(Provided atmospheric pressure is exactly 760 mm of mercury.)

Many things are colder than melting ice. These are given minus temperatures on the scale.

What is temperature?

If your soup is too hot to eat, you don't have to do anything about it. Just wait. As time passes, the particles of the soup lose heat energy to the air. They slow down, and the temperature falls. Eventually, the soup cools to the same temperature as the air. Heat energy no longer flows from soup to air because each soup particle has exactly the same kinetic energy, on average, as each air particle.

When the temperature falls, particles move more slowly.

Two things are at the same temperature if their particles each have the same average kinetic energy.

Same temperature, different heat energy

Don't confuse temperature with heat. The soup in the spoon is at the same temperature as the soup in the bowl. But you would get much less heat energy from it if you accidentally spilled it over yourself.

Absolute zero

Nothing can be cooled below $-273\,°C$. This is the coldest possible temperature. It is called **absolute zero**.

As something cools, its particles move more slowly. They can never stop moving altogether. But at absolute zero, their movement is the minimum possible. They can't go any slower, so the temperature can't fall any more.

The Kelvin scale

Scientists often measure temperatures using the **Kelvin** temperature scale. Its 'degree' is called a **kelvin**, and written **K**. Each kelvin on the scale is the same size as a degree Celsius. But the zero on the scale is at absolute zero.

Converting from one scale to another is easy:

$$\frac{\text{Kelvin}}{\text{temperature}} = \text{Celsius temperature} + 273$$

Temperature:	Celsius		Kelvin
	200 °C	—	473 K
boiling water	100 °C	—	373 K
melting ice	0 °C	—	273 K
	-100 °C	—	173 K
absolute zero	-273 °C	—	0 K

1

−273	0	100	273	373

Say which of these is the temperature of:
a boiling water in °C;   **b** boiling water in K;
c absolute zero in °C;   **d** absolute zero in K;
e melting ice in °C;   **f** melting ice in K.

2 One day in August, these were the temperatures at midday in four holiday resorts:

IBIZA	33 °C
TORREMOLINOS	300 K
PALMA	38 °C
BENIDORM	309 K

Rewrite the list in order of temperature, with the warmest resort at the top.

3

Nick grows tomato plants in a small greenhouse. One summer's day, this is how the temperature in his greenhouse varied, from 9 o'clock in the morning onward:

Time in hours	0	1	2	3	4	5	6	7	8
Temperature in °C	24	28	32	36	39	40	38	36	33

a Plot a graph of temperature against time. Mark the side axis in °C and the bottom axis in hours.
b What was the maximum temperature in the greenhouse in kelvin?
c For how long was the temperature above 36°C?
d Nick meant to open a window when the temperature reached 30°C, to stop his plants getting too hot. But he forgot. At what time should he have opened the window?

2.13 Thermometers

To measure a temperature, you need a thermometer. The problem is deciding which type.

Mercury thermometers

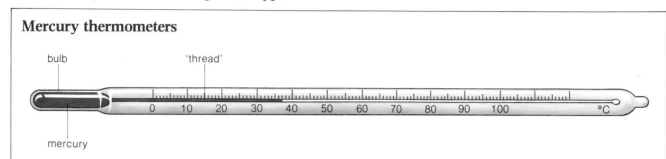

These are made of glass. They have a bulb at one end containing mercury. When the temperature rises, the mercury expands and moves further along the narrow tube.

Mercury thermometers can't measure temperatures below $-39\,°C$ because the mercury goes solid. Most just cover the range from $0\,°C$ to $100\,°C$.

Alcohol thermometers

These work in just the same way as mercury thermometers. But they are cheaper, easier to read, and they can measure lower temperatures. Alcohol doesn't turn solid until $-115\,°C$.
If you have a thermometer in your fridge or freezer, it's probably an alcohol thermometer.

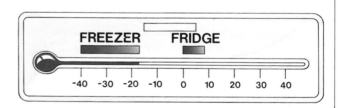

Clinical thermometers

probe contains material which allows more current through when temperature rises

microchip (inside) measures current and converts to temperature reading

battery (inside) supplies current

These are are specially designed to measure the temperature of the human body. Their range is only a few degrees either side of the average body temperature of $37\,°C$. But they measure these temperatures very accurately.

Some clinical thermometers are of the mercury-in-glass type. However, many are electronic like the one above, and have a digital (number) display. There is some more information about electrical thermometers on the next page.

Electrical thermometers

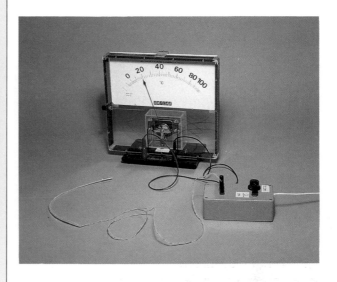

ANSWER:

GO ELECTRICAL!

● **EASY TO READ**
   Choice of pointer reading or digital display

● **READINGS IN COMFORT**
   Scale can be placed well away from the temperature detector

● **CAN BE READ BY COMPUTER**
   For automatic temperature readings, feed the electrical signal straight to a computer.

● **WIDE TEMPERATURE RANGE**

   $-200\,°C$ to $1600\,°C$ or more

A thermocouple thermometer. If one junction is hotter than the other, a current flows. This moves the pointer along the scale.

1 Explain why you probably wouldn't use the following:
a A ordinary mercury thermometer to measure your body temperature.
b A mercury thermometer to measure the temperature of a Bunsen flame.
c A mercury thermometer to measure the temperature of 'dry ice' (about $-80\,°C$).
d A clinical thermometer to measure the temperature at which salted water boils.

2 Say what type of thermometer you would use to measure:
a The temperature of 'dry ice'.
b The temperature of a furnace at around $2000\,°C$.
c The temperature of a kiln around $1000\,°C$.
d The temperature in a freezer.

3 Look at the diagrams on the opposite page. Then write down the temperature reading on:
a the ordinary MERCURY THERMOMETER;
b The ALCOHOL THERMOMETER.

2.14 Expansion

The plane that grows

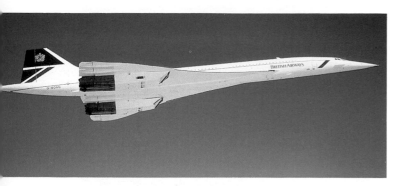

How long is Concorde?
It depends on the temperature.
Like most other materials, the metal body of Concorde expands when heated. At supersonic speeds, when friction from the air heats Concorde's outer skin to around 100°C, the plane is nearly 20 centimetres longer than it is on the ground.

Expansion problems

Materials which **expand** when **heated**, **contract** when **cooled**. Usually the change in size is too small to notice. But it can cause problems. The force produced by an expansion or contraction can be enough to break concrete blocks or buckle steel girders.

Why things expand

Heat a solid, and its particles vibrate more rapidly. The vibrations take up more space. The particles push each other further apart.

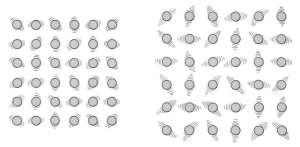

The problem		The solution
On a hot day, concrete runway sections expand. This could cause cracking.	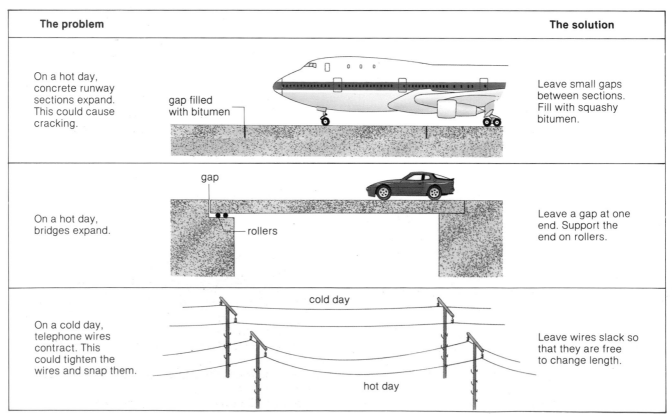 gap filled with bitumen	Leave small gaps between sections. Fill with squashy bitumen.
On a hot day, bridges expand.	gap · rollers	Leave a gap at one end. Support the end on rollers.
On a cold day, telephone wires contract. This could tighten the wires and snap them.	cold day · hot day	Leave wires slack so that they are free to change length.

Expansions compared

Some things expand more than others when heated. The chart shows how much 1 metre lengths of different materials expand when heated by just 100 °C:

Steel rods can be used to reinforce concrete because steel and concrete have the same expansion. If the expansions were different, the steel might crack the concrete on a hot day.

Put an ordinary glass dish straight into a hot oven, and the dish is likely to break. The outside of the glass expands before the inside, and the strain cracks the glass. This shouldn't happen with a Pyrex dish, because Pyrex glass expands much less than ordinary glass.

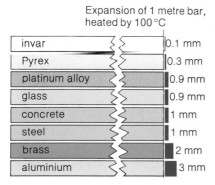

Expansion of 1 metre bar, heated by 100 °C

invar	0.1 mm
Pyrex	0.3 mm
platinum alloy	0.9 mm
glass	0.9 mm
concrete	1 mm
steel	1 mm
brass	2 mm
aluminium	3 mm

Calculating expansion

A **1 metre** bar of steel heated 1°C expands **0.000 01 metres**.
Steel has a **linear expansivity** of **0.000 01/°C**.

If you know the linear expansivity of a material you can work out the expansion of any length for any temperature rise:

expansion = linear × original × temperature
(increase   expansivity length   rise
in length)

For example if the temperature of a 100 m bridge rises by 20 °C on a hot sunny day:

increase = 0.000 01 × 100 × 20 m
in length

= 0.02 m (20 mm)

---

1 Copy, and fill in the blanks:
Most materials __ when heated and __ when cooled. They do so because their __ vibrate more __ and push each other apart. If expansion is resisted it can produce a very high __. To allow for expansion in bridges a __ has to be left at one end.

To answer the following questions, you will need information from the expansion chart on this page.
2 In a light bulb, platinum alloy wires, sealed in glass, are used to carry the electric current to the filament.

Explain why the wires don't crack the glass when the bulb heats up.
3 Expansion can be resisted. Railway lines are fixed to heavy concrete sleepers embedded in chippings, so that they hardly expand at all.
Calculate how much a 1000 metre length of steel railway line would expand, if it were free to, if the temperature rose by 10 °C.
4 An engineer is designing a concrete footbridge over a motorway.
She has to allow for a maximum temperature rise of 40 °C. Has she left enough room for expansion?
40 °C. The linear expansivity of concrete is 0.000 01/°C. Has she left enough room for expansion?

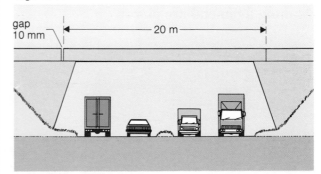

5 Explain why:
a telephone wires are left slack when hung between poles;
b concrete roads have bitumen-filled gaps across them;
c steel can be used to reinforce concrete.

2.15 Using expansion

In home...

Can't get the top off a bottle? Try putting it in hot water. The top expands before the heat reaches the bottle. This makes it a looser fit.

The bimetal strip

This is made by bonding together two thin strips of metal. The one in the diagram is made of brass and invar.

When the strip is heated, the brass expands more than the invar. This makes the strip bend. The brass is on the outside of the bend, because the distance round the outside of the curve is greater than round the inside.

If the strip were cooled, instead of heated, it would bend the opposite way.

Making a bimetal fire alarm

... and factory

Fitting an axle into a wheel.
Liquid nitrogen cools the axle to around $-200\,°C$. This makes it contract, so that it fits easily into the wheel. When the axle warms up again, it expands to give a tight fit.

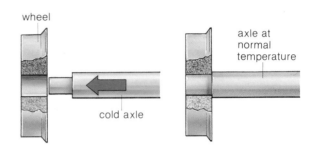

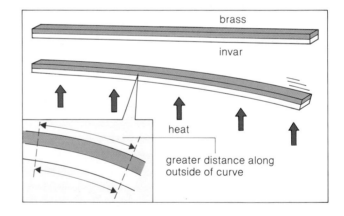

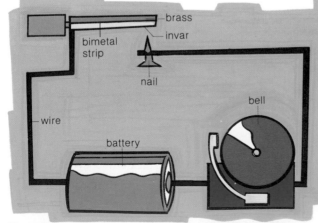

The bell can't work because there is a break in the circuit.

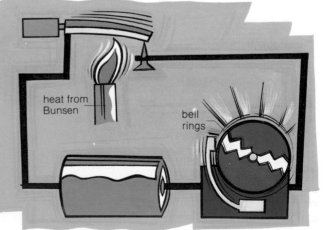

Heat bends the bimetal strip and closes the gap.

Thermostats

There is a **thermostat** fitted to each of these. Its job is to keep the temperature steady:

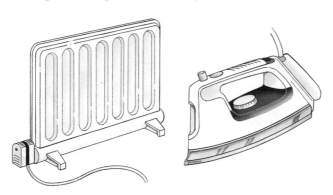

Take the thermostat in the electric room heater. If the room is too hot, it has to cut off the power to the heater. When the room cools down, it has to switch the power back on again. This is done using a small bimetal strip.

When the temperature rises, the bimetal strip starts to bend. Eventually, the contacts separate, and the flow of electricity to the heater is cut. When the temperature falls, the bimetal strip straightens. The contacts touch, and electricity again flows to the heater.

You select the temperature by turning the control knob. Answer question 3 to find out why.

1 Copy and complete:
When heated, brass ___ more than invar.
When cooled, brass ___ more than invar.
If strips of brass and invar are bonded together, they form a ___. This ___ when it is heated.
2 What has happened to each strip to make it bend the way it has?

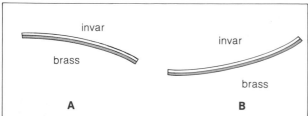

3 Study the diagram of the bimetal thermostat.
a Explain what happens when the temperature in the room rises.
b If the control knob is screwed inwards, the contacts are pushed more firmly together. Will the contacts now separate at a HIGHER or LOWER temperature than before?

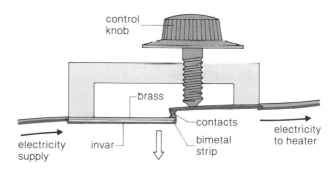

c How would you adjust the control knob to LOWER the temperature in the room?
4 The fire alarm on the opposite page is set off by a Bunsen flame. What adjustment could you make so that it was set off by a much smaller rise in temperature?
5 Over 10 million bimetal strips are at work every day – and that's just in the UK.

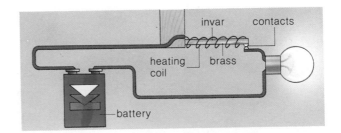

When the light is ON, so is the heating coil.
What effect does this have on the bimetal strip?
What effect does this have on the light and the coil?
What happens to the bimetal strip as a result?
What effect does this have on the light and the coil?
Can you think of a use for this process?

2.16 Expanding liquids

They don't leave space at the tops of lemonade bottles to cheat you. The space is to allow for expansion. Most liquids expand when heated. And they expand much more than solids.

Look under the bonnet of a car to see how much a liquid can expand. A bottle holds the overflow when liquid is pushed out of the cooling system by expansion. The bottle is almost empty when the engine is cold, but nearly full when it is hot.

Using liquid expansion

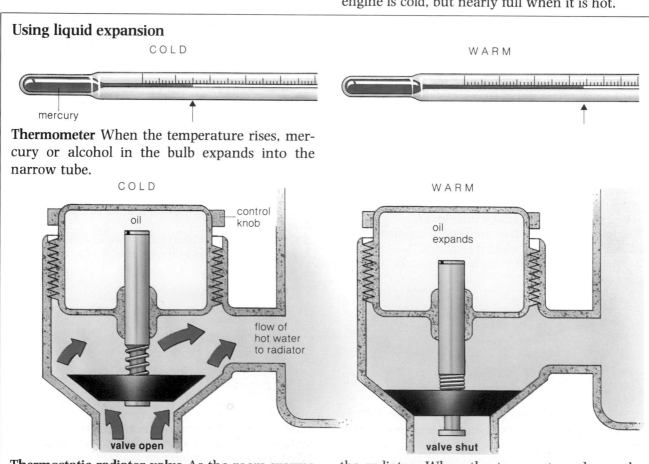

Thermometer When the temperature rises, mercury or alcohol in the bulb expands into the narrow tube.

Thermostatic radiator valve As the room warms up, the oil in the valve expands and pushes the piston down. This shuts off the flow of hot water to the radiator. When the temperature drops, the valve opens again.

Water – the liquid that's different

Water behaves in a very unusual way when heated from 0 °C. As its temperature rises from 0 °C to 4 °C, it actually *contracts*. However, from 4 °C upwards, it expands like any other liquid. This means that water takes up *least space* at 4 °C. It has its *greatest density* at this temperature, and will sink through warmer or colder water around it.

A bitterly cold day.
The puddle freezes but not the lake:

However cold the weather, it takes a very long time for a lake to reach freezing point. As soon as water on the surface cools to 4 °C, away it sinks to the bottom. In fact, none of the water in the lake can start to freeze until all has cooled beneath 4 °C. This can take months of freezing weather.

Not so the puddle. One good overnight frost can cool it all beneath 4 °C. Then freezing soon sets in.

If a lake *does* freeze over, water at the bottom can still be at 4 °C. Fish can survive a severe winter by staying in this deeper, warmer water.

ice

0 °C

1 °C

2 °C

3 °C

4 °C

1 If you are filling a bottle with a drink, why should you leave a space at the top before putting on the cap?
2 Copy the boxes. Write in the FIRST letter of each answer. The result is a word which tells you what happens to water when it is warmed from 0 °C to 4 °C.

?	Abbreviation for Celsius.
?	A thermostatic valve does this when the temperature falls.
?	Does water at 0 °C expand when heated?
?	Device that controls temperature.
?	It warms a room.
?	Liquid used in a thermometer.
?	When a liquid does this, it usually contracts.
?	2 °C is an example.
?	Water at 4 °C does this when surrounded by cooler or warmer water.

3 Explain why fish in a lake can survive a harsh winter, even though the surface of the lake is frozen.
4 Study the diagrams of the thermostatic radiator valve on the opposite page.
Turning the control knob raises or lowers the oil container and piston. This makes the valve shut at a different temperature.
If the oil container and piston are RAISED, will this give you a WARMER room than before? Or a COOLER one?

2.17 Expanding gases

Why shouldn't you . . .

throw aerosols on the fire?

leave a fully inflated dinghy in the Sun

heat an unopened tin of baked beans in the oven?

Gases are squashy. So it is easier to stop a gas expanding than a liquid or a solid. But if a trapped gas isn't allowed to expand, its pressure rises. And the container may not be strong enough to resist it.

Why the pressure rises

In a gas, molecules are always on the move. They travel very fast, hitting each other, and the sides of their container. Struck by billions of molecules, the container feels an outward pressure. If the temperature rises, the molecules move faster. The collisions become more violent; the pressure rises.

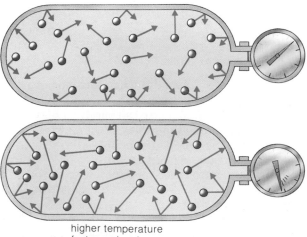

higher temperature
faster molecules
higher pressure

Using gas expansion

If a gas *can* expand, its expansion can be put to work:

In these cylinders, the heat from burning petrol causes more than forty violent gas expansions every second. The force is enough to accelerate the motor cycle to over 50 m/s.

You can use the expansion of gas to undent your table tennis ball.
Dip the ball in very hot water. And then wait. The extra air pressure inside the ball should push out the dent.

Two gas laws

Facts and figures about how temperature affects a gas:

This mass of gas is trapped in a container. Its VOLUME doesn't change.

If its TEMPERATURE is increased, its PRESSURE rises like this:

TEMPERATURE in K	200	400	600	800
PRESSURE in mm of mercury	100	200	300	400

These figures follow some simple rules:

1 If the KELVIN TEMPERATURE doubles, the PRESSURE doubles, and so on.

2 Dividing the PRESSURE by the KELVIN TEMPERATURE gives the same value every time – in this case, 0.5.

Put another way:

The PRESSURE is directly proportional to the KELVIN TEMPERATURE
(provided the VOLUME doesn't change).

This is sometimes called the **pressure law**.

This mass of gas is free to expand. Its PRESSURE doesn't change.

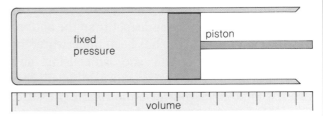

If its TEMPERATURE is increased, its VOLUME rises like this:

TEMPERATURE in K	200	400	600	800
VOLUME in cm³	300	600	900	1200

These figures follow some simple rules:

1 If the KELVIN TEMPERATURE doubles, the VOLUME doubles, and so on.

2 Dividing the VOLUME by the KELVIN TEMPERATURE gives the same value every time – in this case, 1.5.

Put another way:

The VOLUME is directly proportional to the KELVIN TEMPERATURE
(provided the PRESSURE doesn't change).

This is sometimes called the **Charles' law**.

---

1 Explain why:
a a beach ball may burst if left in the Sun;
b air bubbles come out of an empty washing-up liquid bottle if you hold it in hot water without squeezing.
2 Sponge cakes contain thousands of tiny trapped gas bubbles. Explain why a sponge ...

looks like this after it has been left standing for a while:

3 Here is some information about the compressed air in a cylinder, before and after warming:

	Cold cylinder	Warm cylinder
Temperature in K	300	400
Pressure in mm of mercury	600	?
Pressure ÷ Temperature	?	?

Copy the chart and fill in the missing information.

2.18 Squashed gases

Pressure problems

This diver is hunting for pearl oysters. At 30 metres, she is ten times deeper than the bottom of a swimming pool. She has been underwater for over two minutes. And she isn't using any breathing apparatus. The pressure on the diver is around four times that of atmospheric pressure. To survive, she has to expand her lungs to the limit before diving. Even so, on reaching the sea-bed, her lungs are so squashed that the air in them takes up only a quarter of its normal volume.

More squashed air

Pump air into a motor cycle tyre, and it is squashed into about one-third of its normal space. Then, the tyre is full of air at roughly three times the outside pressure. Like all gases, air has a greater pressure when its volume is reduced.

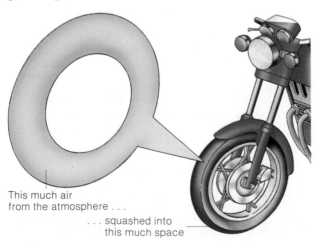

This much air from the atmosphere . . .

. . . squashed into this much space

How pressure and volume are linked

This is the equipment you might use to find the connection between the pressure of a gas and its volume:

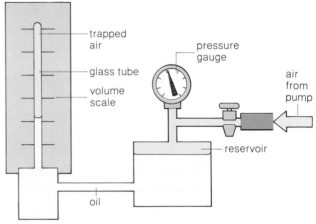

trapped air

glass tube

volume scale

pressure gauge

air from pump

reservoir

oil

The gas is air, trapped above the oil in the glass tube. The volume of the air is reduced in stages by pumping outside air into the reservoir. This forces more oil up the glass tube. Every time the volume is reduced, the pressure of the trapped air is measured on the gauge. Squashing the air warms it slightly, so you have to wait a few moments after each reading for the air to settle to its original temperature.

Here are some typical readings:

VOLUME in cm³	50	40	25	20	10
PRESSURE in mm of mercury	800	1000	1600	2000	4000

There are two connections between the readings:

1 If the VOLUME is HALVED then the PRESSURE is DOUBLED, and so on.

2 Multiplying the PRESSURE by the VOLUME gives the same value every time – in this case, 40 000.

The air obeys **Boyle's law:**
If a fixed mass of gas is kept at a steady temperature, PRESSURE × VOLUME stays the same.

Most other gases obey this law too.

Using Boyle's law

This is the type of problem you could solve using Boyle's law:

A diver working on the sea-bed is breathing out air bubbles. The air in each bubble has a volume of 2 cm³, and a pressure of 3 atmospheres. Up on the surface, the pressure is 1 atmosphere. What is the volume of a bubble when it reaches the surface?

If the temperature doesn't change, PRESSURE × VOLUME stays the same. So,

$$\frac{\text{pressure}}{\text{at 20 m}} \times \frac{\text{volume}}{\text{at 20 m}} = \frac{\text{pressure}}{\text{at surface}} \times \frac{\text{volume}}{\text{at surface}}$$

Filling in the figures,

$$\frac{3}{\text{atm}} \times \frac{2}{\text{cm}^3} = \frac{1}{\text{atm}} \times \frac{\text{volume}}{\text{at surface}}$$

Rearrange this, and the volume of the bubble at the surface works out to be 6 cm³.

Concentrating the molecules

The diagrams show why the pressure rises when a gas is squashed:

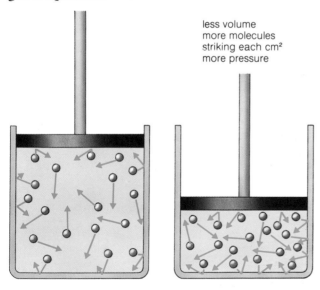

less volume
more molecules
striking each cm²
more pressure

The pressure is caused by fast-moving gas molecules colliding with the sides of the container. If the gas is compressed into a smaller space, the molecules become more concentrated. Each square centimetre of the container sides has more molecules striking it. So the pressure is greater.

A single gas law

It is possible to combine the **pressure law**, **Charles' law**, and **Boyle's law** into a single **gas law**:

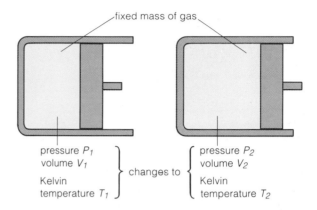

fixed mass of gas

pressure P_1
volume V_1
Kelvin temperature T_1
} changes to {
pressure P_2
volume V_2
Kelvin temperature T_2

If conditions for a fixed mass of gas change so that P_1, V_1, and T_1 become P_2, V_2, and T_2, then:

$$\frac{P_1 V_1}{T_1} = \frac{P_2 V_2}{T_2}$$

Note: when using this equation, the temperatures should be in kelvin (K).

1 According to Boyle's law, if a gas is squashed into a quarter of its original volume, and the temperature does not change, what happens to the pressure?

2 This question is about the experiment and the table of readings on the opposite page.
a If the VOLUME of the gas were only 5 cm³, what would the PRESSURE be?
b Plot a graph of PRESSURE (side axis) against VOLUME (bottom axis).
c From your graph, read off the PRESSURE of the gas when the VOLUME is   i 30 cm³   ii 15 cm³.
d Plot a graph of PRESSURE against 1/VOLUME. (First, use a calculator to work out values of 1/VOLUME.) What shape is your graph?

3 Gas at a PRESSURE of 2 atm is trapped in a plastic container of VOLUME 100 cm³. The TEMPERATURE is 27 °C. In an accident, the container is squashed to only 50 cm³, and heated to 327 °C. As a result, its pressure also changes.
a What is the KELVIN TEMPERATURE before the gas is heated? What is it after the gas is heated?
b Use the information above to make a list showing any values you have been given for P_1, V_1, T_1, P_2, V_2, and T_2. Which of these have you *not* been given?
c Write down the gas law equation. Now rewrite it, substituting the values from your list.
d Work out the new PRESSURE of the gas.
e Use your ideas about particles to explain why the pressure of the gas is more than it was before.

2.19 Conducting heat

Walking on burning coals at 700 °C. Mind over matter? Not necessarily. It may work because coal is a very poor conductor of heat. The firewalker's feet only touch each coal for a short time. So not enough heat flows out to damage the skin. Best not try it yourself however...

How things conduct

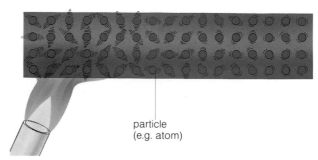

particle
(e.g. atom)

Heat one end of a metal bar, and the particles will vibrate faster. Eventually the particles will vibrate faster all along the bar. Heat has been conducted. The quicker this happens, the better the bar is as a conductor.

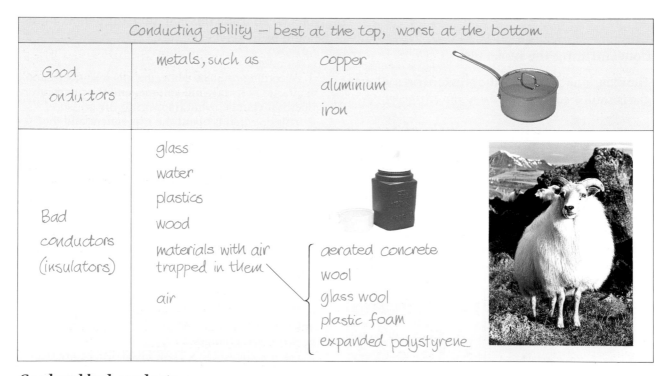

Conducting ability – best at the top, worst at the bottom		
Good conductors	metals, such as	copper
		aluminium
		iron
Bad conductors (insulators)	glass	
	water	
	plastics	
	wood	
	materials with air trapped in them	aerated concrete
		wool
		glass wool
		plastic foam
	air	expanded polystyrene

Good and bad conductors

Metals are the best conductors of heat. Non-metals like wood and plastic are bad conductors. So are most liquids. Gases are worst of all. You can sometimes tell how well something conducts just by touching it. A metal door handle feels cold because it quickly conducts heat away from your hand. A polystyrene tile feels warm because it hardly conducts away any heat at all.

Poor conductors of heat are called **insulators**. Many materials are good insulators because they have tiny pockets of air trapped in them.

Insulating the house

In a house, good insulation means lower fuel bills. Here are some of the ways in which insulating materials are used to cut down heat loss:

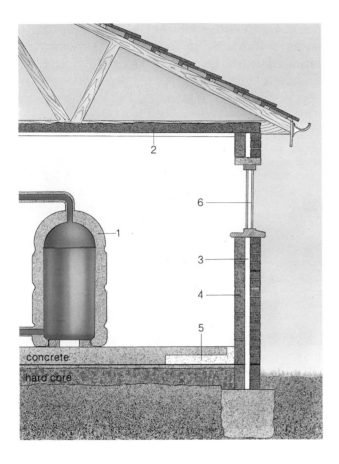

concrete
hard core

1 Plastic foam lagging round the hot water tank.
2 Glass wool insulation in the loft.
3 Air cavity between the inner and outer walls.
4 Inner wall built from highly insulating aerated concrete blocks. The concrete has tiny bubbles of air trapped in it.
5 Polystyrene insulation under the edge of the floor.
6 Double-glazed windows. Two sheets of glass, with an insulating layer of air between them.

1 Explain why:
a an aluminium window frame feels cold when you touch it, but a wooden frame feels warm;
b aerated concrete is a better insulator than normal concrete;
c it is much safer picking up hot dishes with a dry tea-towel than with a wet one;

U-values

To calculate likely heat losses from a house, architects need to know the **U-values** of different materials. For example:

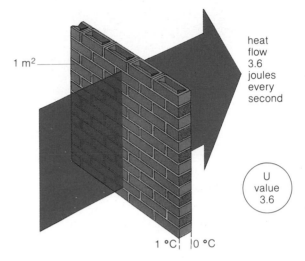

1 m²

heat flow 3.6 joules every second

U value 3.6

1 °C   0 °C

A single-brick wall has a U-value of 3.6 W/(m² °C). This means that the 1 square metre wall, with a 1 °C temperature difference across it, will conduct heat at the rate of 3.6 joules every second.

The heat flow would be greater if:
a the temperature difference were higher;
b the area was greater;
c the wall was thinner.

This is how the U-values of different materials compare. The lower the U-value, the better the insulation.	
	U-value: W/(m² °C)
Single-brick wall	3.6
Double wall, with air cavity	1.7
Double wall, with insulating foam in cavity	0.5
Glass window, single layer	5.7
Double-glazed window	2.7

d Dave feels cold in his string vest, but warm if he puts a tight shirt over it.
2 Look at the table of U-values on this page. Then explain why houses with small, single-layer windows are likely to lose less heat than those with larger windows. Is this true if the windows are double-glazed and the wall cavities are insulated?

2.20 Convection

Unless the pilot can find a thermal, this glider isn't likely to stay airborne for very long. Thermals are rising currents of warm air. They can occur above hilltops, over factories, or under clouds. The problem is finding them. Experienced pilots keep a look-out for circling birds. Birds discovered the secrets of thermal soaring many millions of years before people.

Whenever warm air rises, cooler air moves in to take its place. The result is a circulating current of air called a **convection current**. Convection doesn't only happen in air. It can occur in all gases and liquids.

Two simple experiments show convection in air and in water:

Hot air rises above the candle. Cooler air flows in to replace it. The smoke from the burning straw shows the current of air.

Hot water rises above the Bunsen flame. Cooler water flows in to replace it. Potassium permanganate crystals colour the water so that you can see the current.

card

purple crystals

low flame

Why warm air rises

When air is heated it expands. This makes it less dense because the same mass now takes up a larger volume. Being less dense, the warm air floats upwards through the denser, cooler air around it.

Weather convection

In a cloud like this, warm damp air can rise at speeds of 30 metres per second or more. As the air rises, it cools. This forms new cloud. Read page 103 to find out why.

Cutting convection

It works for eggs as well as heads. Put an insulated cover over any warm surface, and the circulation of air is cut down. Also, the outside air isn't heated so much. Either way, the heat lost by convection is now reduced.

Using convection

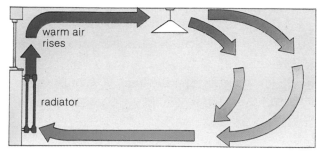

warm air rises

radiator

Most of the heat from a radiator is circulated by convection. Warm air rises above the radiator. It carries heat all round the room.

In a fridge, cold air sinks below the freezer compartment. This sets up a circulation which cools all the food in the fridge.

freezer compartment

cold air sinks

main supply

overflow pipe to allow for expansion

header tank

storage tank

hot water rises

boiler

Supplying taps with hot water –
In the simple system above, water is heated in the boiler. It rises to the storage tank. Cooler water flows in to replace it. It too is heated. In time, a supply of hot water collects in the tank from the top down. The header tank provides the pressure to push the hot water out of the taps.

1 Explain why:
a The freezer compartment in a fridge is placed at the top.
b A fridge doesn't work properly if the food is too tightly packed inside.
c A radiator quickly warms all the air in a room, even though air is a very poor conductor of heat.
d Warm water rises when surrounded by cooler water
2 This person feels a draught when the bonfire burns fiercely. Why?

Draw a diagram to show the flow of air.
3 Two freshly poured cups of hot tea:

One is covered, the other isn't.
This is how the temperature of each drops with time:

TIME in minutes	0	2	4	6	8	10
TEMPERATURE (covered) in °C	80	73	66	61	56	52
TEMPERATURE (uncovered) in °C	80	67	56	48	42	37

a Plot a graph of TEMPERATURE (side axis) against TIME (bottom axis) for each cup. *Use the same axes for both graphs.*
b What is the difference in temperature between the two cups after 5 minutes?
c Most people don't like drinking tea which is cooler than 45 °C. Estimate to the nearest half minute how long you could leave the uncovered drink before drinking it.

2.21 Holding heat

June, Newquay. The beach may be warm. But the sea most definitely isn't. In fact, it will take so much heat energy to warm up the sea, that it won't reach its maximum temperature until the end of September. However, there's good news for winter swimmers. With so much heat absorbed, it will then take the sea a long time to cool down again. Chances are that, in January, you'll find the sea warmer than the beach.

Specific heat capacity

It takes 4200 joules of heat energy to heat 1 kilogram of water just through 1°C:

water has a specific heat capacity of 4200 joules per kilogram per °C (written 4200 J/kg °C).

This is how water compares with other materials:

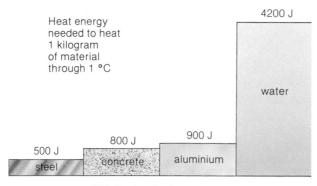

This is also the heat energy given **out** when 1 kilogram **cools** by 1 °C

Calculating heat energy

For water:

       4200 J heats 1 kg through 1 °C
So,   8400 J heats 2 kg through 1 °C
So, 84 000 J heats 2 kg through 10 °C

You could have calculated these results using an equation:

heat energy gained	=	mass	×	specific heat capacity	×	temperature rise
(J)		(kg)		(J/kg °C)		(°C)

The equation works for other things as well as water. You can also use it to calculate the heat energy lost when the temperature falls.

Storing heat

Water has a high specific heat capacity. This makes it a very useful material for storing and carrying heat energy:

A hot-water bottle can keep your feet warm for an hour or so.

Water carries heat from the boiler to the radiators around the house.

Water carries unwanted heat from the engine of a car to the radiator.

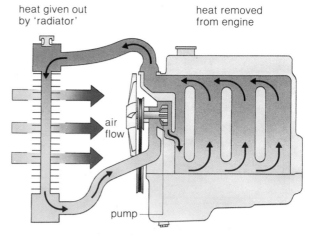

Night storage heaters use concrete blocks to store heat. Concrete doesn't have as high a specific heat capacity as water. But it is more dense, so the same mass takes up less space. Heating elements warm the blocks overnight when electricity is cheap. The blocks give out heat through the day as they cool down.

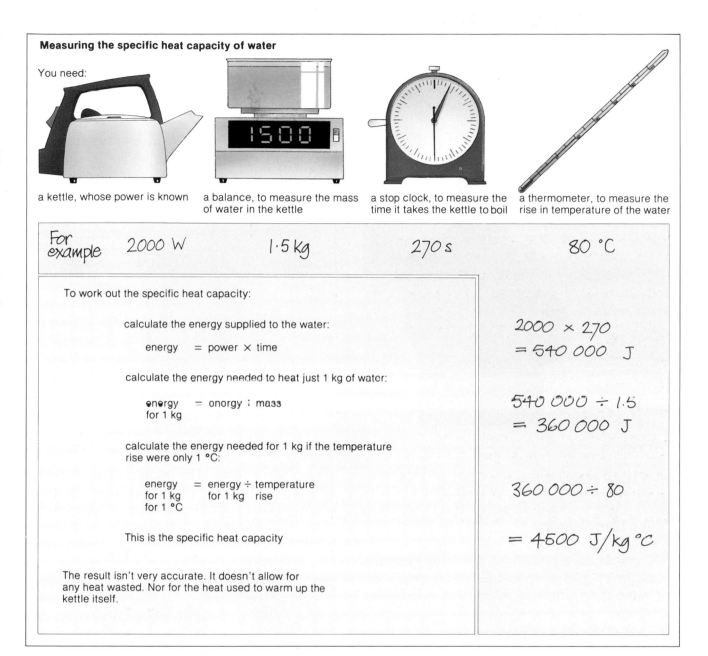

Measuring the specific heat capacity of water

You need:

a kettle, whose power is known

a balance, to measure the mass of water in the kettle

a stop clock, to measure the time it takes the kettle to boil

a thermometer, to measure the rise in temperature of the water

For example   2000 W   1·5 kg   270 s   80 °C

To work out the specific heat capacity:

calculate the energy supplied to the water:

energy   = power × time

$2000 × 270$
$= 540\ 000\ J$

calculate the energy needed to heat just 1 kg of water:

energy   = energy ÷ mass
for 1 kg

$540\ 000 ÷ 1.5$
$= 360\ 000\ J$

calculate the energy needed for 1 kg if the temperature rise were only 1 °C:

energy   = energy   ÷ temperature
for 1 kg   for 1 kg   rise
for 1 °C

$360\ 000 ÷ 80$

This is the specific heat capacity

$= 4500\ J/kg\ °C$

The result isn't very accurate. It doesn't allow for any heat wasted. Nor for the heat used to warm up the kettle itself.

1 If you eat a jam tart straight from the oven, the jam is at the same temperature as the pastry, but it's more likely to burn your tongue:
● Nick says this is because there's a greater mass of jam than pastry.
● Deborah says it's because the jam has a higher specific heat capacity than the pastry.
● Razia says that it's a combination of both reasons. Who is right?
2 Steve has a bright idea. Instead of filling his hot-water bottle with water, he will fill it with hot air from the hairdryer. Then he won't have to worry about leaks.

Here is some information about the water or air he could use in the bottle.

	Water	Air
Mass in kg	0.5	0.0005
Specific heat capacity J/kg °C	4200	1000
Temperature in °C	75	75

a Calculate the heat given out when a water-filled bottle cools to 25 °C.
b Explain why you think that Steve's idea is, or is not, a good one.

2.22 Coping with cold

Cold hands. But at least the core of his body is warm. And if his temperature drops beneath the normal 37 °C, then his automatic temperature control system goes into action:

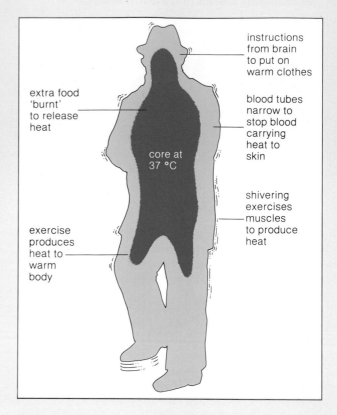

instructions from brain to put on warm clothes

extra food 'burnt' to release heat

blood tubes narrow to stop blood carrying heat to skin

core at 37 °C

shivering exercises muscles to produce heat

exercise produces heat to warm body

At risk

If the body loses too much heat, the core temperature starts to drop. If it drops more than 2 °C, the body stops working properly. The condition is called **hypothermia**.

Old people are especially at risk from hypothermia. Every winter, thousands of old people die because they cannot afford to heat their homes properly.

Young babies also find it difficult to cope with the cold. They don't store as much body heat as adults, so a loss of heat can have a more drastic effect. And they can't adjust to sudden heat losses because their temperature control system isn't fully developed.

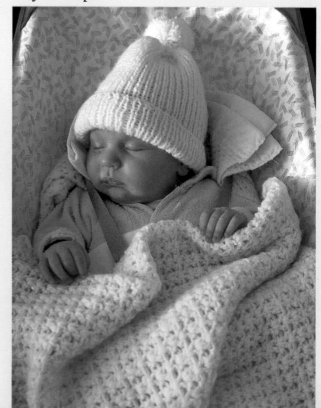

Survival on the hills

Early March. A mild sunny day suddenly turns cold wet and windy. Down in the town, the weather is just a nuisance. But for walkers up on the hills, it can be a killer.

Water evaporating from wet clothes quickly takes heat from the body. And a sharp wind makes matters worse. A 30 mph wind has the same chilling effect as a 40 °C drop in air temperature. Accidents bring extra problems, because the risk of hypothermia ('exposure') is much greater if someone is injured and can't move.

To face the weather, hill walkers need to be properly equipped:

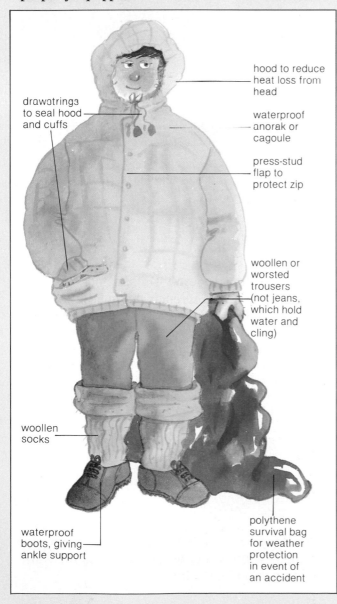

drawstrings to seal hood and cuffs

hood to reduce heat loss from head

waterproof anorak or cagoule

press-stud flap to protect zip

woollen or worsted trousers (not jeans, which hold water and cling)

woollen socks

waterproof boots, giving ankle support

polythene survival bag for weather protection in event of an accident

Survival at sea

Without an insulating suit this North Sea diver would be suffering from hypothermia in a matter of minutes. In a cold sea, the human body loses heat over 20 times faster than in cold air.

The wet-suit is worn with a layer of water trapped between the suit and the diver's skin. The water is an insulator. More insulation comes from the thousands of tiny nitrogen bubbles trapped in the suit lining.

Can you think of reasons why:

● babies store less body heat than adults?

● old people are more at risk from hypothermia than younger people?

● hill walkers and climbers are more at risk from hypothermia if they are injured?

● a swimming pool changing room is kept at a temperature of 25 °C, but the pool itself is kept at 30 °C?

2.23 Melting and freezing

Ice is a marvellous substance for keeping things cool. Not just because it is cold – but because it absorbs so much heat when it melts.

The quick way to cool a drink. Just add a couple of ice cubes and stir. As the ice melts, it absorbs so much heat from the drink that the temperature falls to near-freezing.

Ice pack in a cool-box. As the ice melts, it absorbs so much heat that the food stays as cool as the fridge, for 24 hours or more.

Arctic icebergs have been found drifting as far south as Florida. It would take more than 10 000 000 000 000 joules of heat energy to completely melt this iceberg.

Latent heat

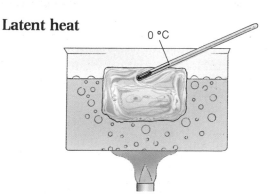

0 °C

Putting heat into ice at 0 °C.
The ice melts. But it doesn't get any hotter. The temperature stays at 0 °C until all the ice has melted.

The heat absorbed when a solid melts is called **latent heat of fusion** (*fusion* means *melting*; *latent* means *hidden*).
The effect of the heat seems to be hidden because

the temperature doesn't rise. In fact, the heat absorbed is used to wrench the molecules of the solid apart, so that they are free to move around as a liquid.

For ice to melt, each kilogram has to absorb 330 000 joules of heat energy:
water has a **specific latent heat of fusion** of 330 000 joules per kilogram.

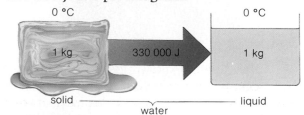

0 °C — 1 kg — solid — 330 000 J — 0 °C — 1 kg — liquid — water

This means that it takes nearly as much energy to change ice into water as it does to heat the water right up to boiling point.

Volume changes

When it melts,

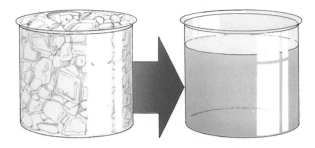

this much ice . . . . becomes this much water.
Water takes up less space as a liquid than as a
solid.
This is because of the way the molecules stick
together. In ice the molecules are grouped in
rings. This takes up a lot of space. When the ice
melts, the rings are broken. So the molecules can
pack together more closely.

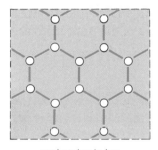

molecules in ice

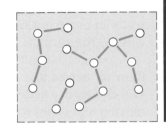

molecules in liquid water

When water changes to ice, it expands. If the ice is
trapped, the expansion produces a huge force.
This can:
– burst pipes
– lift paving stones
– crack stonework off buildings
– split pieces of rock from a rock face.

Lowering the freezing point

Adding salt to water makes the freezing point go
down. Damp roads which would be completely
frozen over at $-10\,°C$, will stay clear of ice if salt is
sprinkled over them.

Adding antifreeze to water keeps it liquid down to
$-25\,°C$ or lower.

Finding a melting point

Melting point means the same as **freezing point**. It
is the temperature at which a solid turns liquid, or
a liquid turns solid. This is how you can find the
melting point of wax:

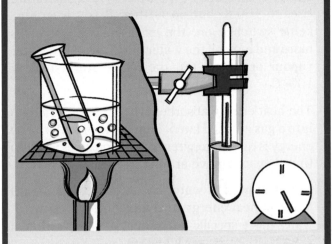

Put some candle wax in a test-tube.
Melt the wax by putting the tube in boiling water.
Let the melted wax cool.
Measure the temperature every minute.
While the wax is liquid, the temperature falls;
while the wax is solid, the temperature falls.
But while the wax is changing from liquid to solid,
the temperature stays the same.
This temperature is the melting point.

1 Explain why, in winter:
a A car engine may be damaged if antifreeze isn't
added to the water in its cooling system.
b Heaps of snow may still be standing at the roadside
days after the weather has become warmer.
2 Water has a specific latent heat of fusion of
330 000 J/kg. How much heat energy is needed to
melt:
1 kg of ice?   2 kg of ice?   10 kg of ice?
3 This is how the temperature of some melted wax
changed when it was left to cool:

TIME in minutes	0	1	2	3	4	5	6	7	8
TEMPERATURE in °C	83	70	60	53	53	53	53	48	43

a Plot a graph of TEMPERATURE against TIME.
b How long was it before the wax started to turn
solid?
c How long was it before the wax had all turned
solid?
d What is the melting point of the wax?

2.24 Making vapour

It happens every time you put the kettle on. Heat energy is absorbed by the water. The temperature rises to 100 °C but no further. If you leave the kettle switched on, the extra energy just turns more and more of the water into a gas called **water vapour** or **steam**. But the temperature stays at 100 °C.

The heat energy absorbed when a liquid changes into a gas is called **latent heat of vaporization**. The energy is needed to wrench the molecules apart so that they can move around freely as a gas.

Each kilogram of water has to absorb 2 300 000 joules of heat energy to change into vapour: water has a **specific latent heat of vaporization** of 2 300 000 joules per kilogram.

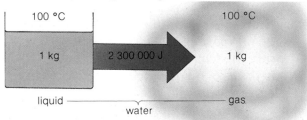

Compare this with the specific latent heat of fusion of water on page 98. It takes around seven times as much heat energy to change water into vapour as it does to change ice into water.

Getting back the latent heat.
Jet of steam meets cold coffee.
Steam condenses – it changes back to a liquid.
Result: large release of heat. Hot coffee in next to no time.

Evaporation

A liquid doesn't have to boil to change into a gas. Even on a cool day, rain puddles can vanish and wet clothes can dry out.

If a liquid is changing into a gas, it is **evaporating**. Liquids evaporate because some of their molecules move faster than others. Most of the molecules stick together. But faster ones close to the surface may escape from the others and form a gas.

How to make a liquid evaporate more quickly:

Increase the temperature
Wet clothes dry better on a warm day because more of the water molecules have enough energy to escape.

Increase the surface area
Water in a puddle dries out more quickly than water in a cup because more of the molecules are close to the surface.

Blow air across it
Wet clothes dry better on a windy day because the moving air carries escaping molecules away more quickly.

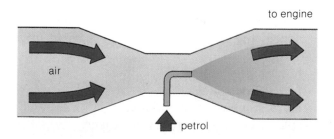

Turn it into a spray
A spray is made up of millions of tiny liquid droplets. It gives the liquid a much larger surface area. The idea is used in the carburettor fitted to some engines. Petrol evaporates as it is sprayed into a moving stream of air. The mixture of air and petrol vapour is sucked into the engine to be burnt.

Boiling

Boiling is a very rapid form of evaporation. When water boils, you can see bubbles forming deep in the liquid. They grow, rise, and burst from the surface, carrying large amounts of vapour with them.

Even a cold liquid has tiny vapour bubbles inside it. But these are squashed by the pressure of the atmosphere pushing on the water. As the temperature rises, the pressure inside the bubbles grows stronger. Eventually, it is strong enough to overcome atmospheric pressure. The bubbles grow quickly. The liquid boils.

A lower boiling point

At the top of Mount Everest, water boils at only 70 °C. Atmospheric pressure is much lower than at sea-level. So the vapour bubbles don't have to be so hot to overcome the pressure.

A higher boiling point

In a pressure cooker, water boils at about 120 °C. Trapped steam puts the pressure up, and this raises the boiling point of the water. The higher temperature means faster cooking.

Adding plenty of salt to water raises the boiling point by several degrees. It may mean slightly faster cooking, but it isn't good for the taste or your health.

---

1 [?] A gas becomes this when it condenses.

[?] The surface of a puddle of water has more of this than a beaker of water.

[?] If this rises, water becomes vapour more quickly.

[?] A liquid does this when it becomes a gas.

[?] Does water have to boil to become vapour?

[?] This stays the same when water is boiling.

Copy the boxes. Fill in the first letter of each answer to make a word. This is the heat absorbed when a liquid becomes a gas.

2 Explain why:
a food cooks faster in a pressure cooker;
b once the water in a kettle has boiled, it doesn't go on getting hotter and hotter, even if the kettle is left switched on.

3 A kettle has a power output of 2300 W. It is full of boiling water.
The specific latent heat of vaporization of water is 2 300 000 J/kg.
a How much heat energy is needed to change 1 kg of water into steam?
b How much heat energy is put into the water every second?
c How long will it take the kettle to produce 1 kg of steam?

4 Mike and Janet have used a kettle to boil drinking water while climbing in the Alps, sea water on the beach, and tap water at home. They have measured the three temperatures: 100 °C, 95 °C, and 103 °C. But they can't remember which is which. Decide for Mike and Janet.

2.25 Cooling by evaporation

Evaporation has a cooling effect.
You only have to stand around in wet clothes to discover that. The water has to absorb latent heat to evaporate. So when your clothes dry out, heat is taken from your body. If there is a wind blowing, the water evaporates faster, and you chill even more.

Sometimes, your body needs to lose heat. Sweating does this for you automatically. You start to sweat if your body temperature rises more than $\frac{1}{2}$ °C above normal. The sweat comes out of tiny holes in your skin. As it evaporates, it takes heat away from your body.

Dogs don't sweat through their skin. To lose heat, they hang out their tongues and pant. As moisture evaporates from the tongue, it takes heat away from the bloodstream. The panting speeds up evaporation because it blows air across the tongue.

Cool in the fridge

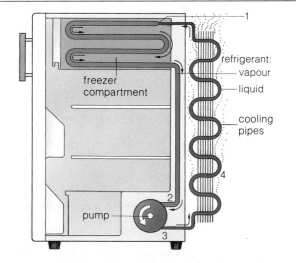

A refrigerator takes heat from the food inside, and gives it out at the back. Put your hand over the back and you can feel the warm air rising.

Evaporation keeps food cool in the fridge.
In their pipework, refrigerators have a substance called a **refrigerant**. This evaporates easily.

Follow the refrigerant round the circuit:

1. Liquid evaporates in pipes in freezer compartment. Heat taken from food and air in fridge.
2. Vapour drawn away by electric pump.
3. Electric pump compresses vapour. Vapour turns liquid. Latent heat is released.
4. Hot liquid forced through cooling pipes. Heat given off.

Vapour in the air

On 'close' or **humid** days, the air is so full of water vapour that other water evaporates very slowly. Sweat tends to stay on your skin. There is hardly any cooling effect. So you feel hot and uncomfortable.

Warm air can hold more water vapour than cold air.

If warm humid air suddenly cools, some of its vapour has to condense.

It may become billions of tiny water droplets in the air. You see these as white cloud, steam or mist. It may become condensation on any cold surface. If condensation freezes, the result is frost.

These all have something in common:

clouds of steam from a kettle...

frost...

condensation on windows...

fog, mist, and clouds...

They are all formed when humid air cools.

1

evaporation	latent heat	condensation
water vapour	humidity	clouds

Choose one item from the table to match each of the following:
a Warm air can hold more than cold air.
b Formed when water vapour meets a cold surface.
c Absorbed when a liquid evaporates.
d Millions of tiny water droplets in the air.

2 Explain why:
a if you tip white spirit on the back of your hand, the spirit vanishes and your hand feels cold;
b on a humid day, you feel hot and uncomfortable;
c ... but you don't feel so uncomfortable if there is a breeze blowing.

3 In a refrigerator:
a what happens to the refrigerant when it absorbs latent heat?
b where does it absorb latent heat?
c what happens to the Freon when it releases its latent heat?
d where does it release its latent heat?

4 Sue goes to check her car one morning. She finds that there is condensation on the INSIDE surface of the windows.
Say whether each of the following is TRUE, or FALSE, or whether you CAN'T TELL without more information.
A It is cooler outside the car than inside.
B It is cooler inside the car than outside.
C There is humid air outside the car.
D There is humid air inside the car.

2.26 Changing the weather

Summer all the year round.
This weather is made by technicians.
Outside, the weather happens naturally.
But it can still be changed by human activity.

Living in a heat island

Towns are warmer than the country. And less windy.
But they get less sunshine and more rain.
Here are some of the reasons:

- Towns produce heat. It comes from factories, vehicles and heating systems.

- Towns have huge surfaces to absorb the Sun's heat. And they are full of materials like bricks, stone and concrete which are good at storing heat.

- Smoke and exhaust gases from vehicles collect above towns and act as a huge insulating blanket.

- Towns are windbreaks. They slow the air, so that heat isn't carried away. Each town becomes an 'island' of heat. As warm air rises from it, clouds and thunderstorms may develop.

Do-it-yourself weather

Gardeners create their own 'microclimates' using greenhouses, cloches, and windbreaks.

Central London gets almost twice as much thundery rain as its suburbs. On a summer's evening, it can be as much as 10 °C warmer. And it has at least one more frost-free month every year.

Too hot to bear

Heat and humidity can make life very uncomfortable. This chart shows the 'zone of comfort' which suits people best:

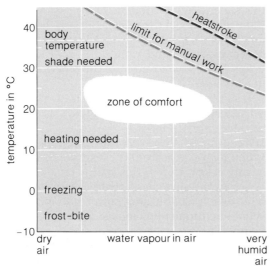

Many people think that city violence comes from frustration, unemployment, and poor housing. And statistics show that it is most likely to erupt on hot summer evenings when the build-up of heat in an overcrowded city can make life unbearable.

Outlook: warmer, but wetter

That's the world outlook for the next few centuries. But it isn't good news. The tropics will be wetter, but countries like Ethiopia and Sudan, which already have severe drought problems, are likely to be even drier.

The main cause of the change is the slow build-up of carbon dioxide gas in the atmosphere. All living things give off carbon dioxide. Plants absorb it. But extra carbon dioxide is pouring out of power-stations, homes and vehicles as fuels are burned. And that is upsetting the balance, because plants can't absorb the extra.

Carbon dioxide in the air is like the glass in a greenhouse. It lets the Sun's radiated heat reach the ground, but traps heat radiated from the ground. Scientists expect the **greenhouse effect** to lift world temperatures by about 4 °C over the next few hundred years. It isn't much. But it will be enough to change the world's climate. How we use fuels today is going to affect the lives of many generations in the future.

Look at the 'zone of comfort' chart on this page. What is the maximum comfortable temperature for manual work when the air is very humid? How does this change if the air is drier? Can you explain why?

This map shows a block of houses and the different heat-absorbing materials around. Draw a map like this for your home and its surroundings. Mark on anything that acts as a windbreak. Then mark on the most windy places, the most sheltered places, and the places you would choose if you were going to sunbathe.

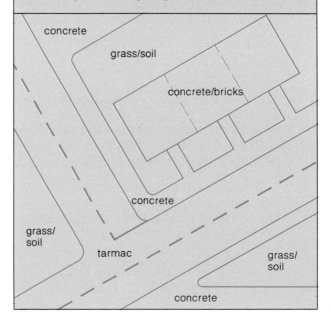

Questions on Section 2

1 The diagram below shows a model power-station. A small steam engine drives a generator which lights a bulb. Decide where each of the following energy changes is taking place. (You can answer by writing one of the letters A-D in each case.)

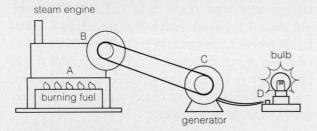

a Kinetic energy to electrical energy.
b Heat energy to kinetic energy.
c Electrical energy to heat and light energy.
d Chemical energy to heat energy.

2 The diagram below shows a hydroelectric scheme. Water rushes down from the top of the lake to the power-station. In the power-station, the water turns a turbine which drives a generator.

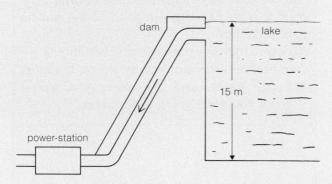

a Why is the dam thicker at the base than at the top?
b What type of energy does the water have when it reaches the power-station?
c Some of the water's energy is wasted.
   i Why is energy wasted?
   ii What happens to the wasted energy?
d What is the gravitational potential energy of 1 kg of water at the top of the lake? (Assume $g = 10 \, \text{N/kg}$.)

e If 1000 kg of water flows from the lake every second, how much potential energy is lost by this water every second?
f If the efficiency of the scheme is 50%, how much electrical energy does the power-station deliver every second?
g What is the power output of the power-station in kilowatts?

3 The hydroelectric scheme in question 2 is a renewable energy source.
a What is meant by a *renewable energy source*?
b When water flows from the lake, potential energy is lost. How is this energy replaced?
c What advantages does a hydroelectric scheme have over a fuel-burning power-station?
d What environmental damage does a hydroelectric scheme cause?

4 The diagram below shows a working model of a hydraulic jack.

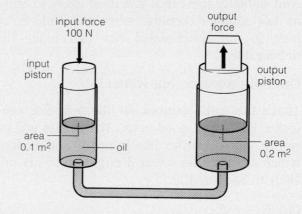

a Why does the output piston rise when the input piston is pushed down?
b Why does the output piston move a shorter distance than the input piston?
c What is the oil pressure immediately under the input piston?
d What is the oil pressure immediately under the output piston?
e What is the upward force on the output piston?
f Is this jack a *force magnifier* or a *movement magnifier*? Give a reason for your answer.
g What change would you make to the hydraulic jack in order to produce a higher output force for the same input force?

5 The diagram below shows a refrigerator. In and around a refrigerator, heat is transferred by conduction, by convection, and by evaporation. Decide which process is mainly responsible for the heat transfer in each of the following:

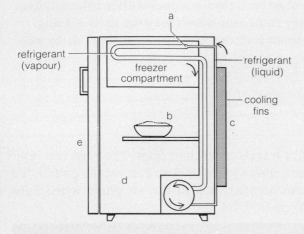

a Heat is absorbed as liquid refrigerant changes to vapour in the pipework.

b Cool air descending from the freezer compartment takes away heat from the food.

c Heat is lost to the outside air through the cooling fins at the back.

d Some heat from the kitchen enters the refrigerator through its outer panels.

e Some heat enters the refrigerator every time the door is opened.

6 The diagram below shows a hot-water storage tank. The water is heated by an electric immersion heater at the bottom.

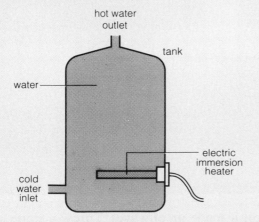

a How could heat loss from the tank be reduced? What materials would be suitable for the job?

b Why is the heater placed at the bottom of the tank rather than the top?

c The heater has a power output of 3 kW.
  i What does the 'k' stand for in 'kW'?
  ii How many joules of heat energy does the heater deliver in one second?
  iii How many joules of heat energy does the heater deliver in 7 minutes?

d The tank holds 100 kg of water. The specific heat capacity of water is 4200 J/(kg °C).
  i How much heat (in joules) is needed to raise the temperature of 1 kg of water by 1 °C?
  ii How much heat (in joules) is needed to raise the average temperature of all the water in the tank by 1 °C?
  iii If the heater is switched on for 7 minutes, what is the average rise in temperature of the water in the tank (assuming that no heat is lost)?

7 On deep-sea dives, divers sometimes use a diving bell like the one below. The bell contains a pocket of air whose pressure is the same as that of the water outside. The divers can enter or leave the bell through the hole in the bottom.

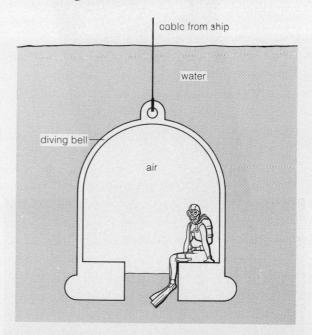

At the surface, air pressure is 100 kPa. As a diving bell descends into the water, the pressure on it increases by 100 kPa for every 10 m of depth.

a Make a table showing the pressure on the diver at depths of 0 m, 10 m, 20 m, and 30 m.

b At what depth is the pressure three times that on the surface?

c When at the surface, the diving bell holds 6 m³ of air. If the bell is lowered to a depth of 20 m, and no more air is pumped into it, what will be the volume of the trapped air? (Assume that the temperature of the trapped air does not change.)

3.1 Rays of light

Almost anything can give out light.

You see some things because they give off their own light: the Sun, for example.

You see other things because daylight, or other light, bounces off them. They **reflect** light into your eyes: this page, for example.

Rays and beams

In diagrams, **rays** are lines with arrows on them. They show you which way the light is going.
A **beam** is drawn using several rays side by side.

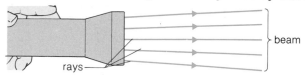

Some facts about light

Light carries energy. This calculator doesn't need a battery to keep it going. Just plenty of sunlight.

Light waves can travel through empty space. Otherwise it wouldn't be possible to see the Sun and the stars. The waves travels very fast – about 300 000 kilometres every second.

Light travels in straight lines. The edge of a laser beam shows you this. You can see the path of the beam because dust in the air glints when light reflects from it.

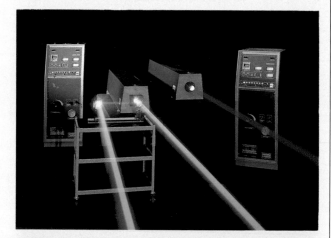

Light is made up of waves. Throw a stone into a pond and ripples spread across the surface of the water. Light travels in much the same way. But the 'ripples' are tiny electric and magnetic vibrations. And they don't need water to travel across.

Holograms

The picture on the credit card is called a **hologram**. It looks three-dimensional. And its colour changes as you look at it from different angles. The image is actually a pattern of light waves reflecting from the surface of the card. The hologram is put on the card to make it difficult for a forger to copy.

Lasers

Lasers give out a very intense beam of light. The beam is extremely narrow. It is just one colour.

1 Which of the following give off their own light?
a page of a book;
b the Sun;
c traffic lights.
2 Write down *three* uses of lasers.

Surgeons use laser in delicate operations on eyes and nerves. The fine beam gives a concentrated heat which can seal blood vessels and cut tissue very accurately.

Lasers are used in compact disc (CD) players (see page 135). As the disc rotates, light from a tiny laser is reflected in pulses from it. The pulses are changed into electrical signals and then into sound. Telephone systems also use pulses of laser light to send speech long distances through optical fibres (see page 227).

At many supermarket check-outs, the price of each item is 'read' by passing its bar-code over a laser. A detector picks up the reflected beam. The light pulses are changed into electrical signals for the till.

3 Pia isn't convinced that light is a form of energy. She wants evidence for this as well. What can you suggest?
4 What is the speed of light through space?

3.2 Flat mirrors

The door isn't as smooth as the mirror. It scatters light in every direction.

Mirrors reflect light in such a way that they produce images.

Laws of reflection

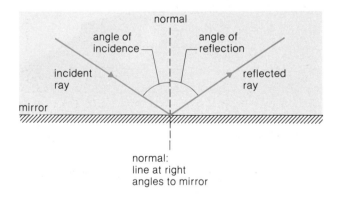

normal:
line at right
angles to mirror

When a ray of light is reflected from a mirror, it obeys two simple rules:

1 The angle of reflection is equal to the angle of incidence. The ray is reflected from the mirror at the same angle as it arrives.

2 The ray striking the mirror, the reflected ray, and the normal all lie in the same plane. You can draw all three on one flat piece of paper.

These are the **laws of reflection**.

How a flat mirror makes an image

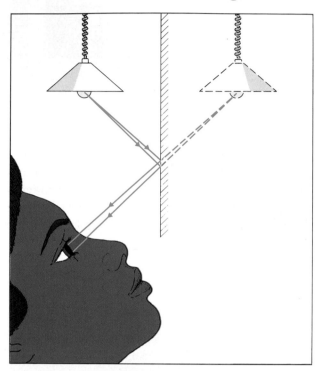

Thousands of rays could be drawn from the lamp. But to keep things simple, only two are shown. The rays are reflected into the eye. They seem to come from a position behind the mirror. This is where you would see an image of the lamp.

Rays don't actually pass through the image. They just *seem* to come from it. The image is called a **virtual image**. It can't be put on a screen.

More rules

When something is put in front of a flat mirror, its image is:

the same size;

the same distance from the mirror, and in a matching position;

laterally inverted – 'left' becomes 'right' and 'right' becomes 'left'.

Finding the image

You can find the position of an image by experiment:

1 Stand a mirror along the middle of a piece of paper. Draw a pencil line along the back of the mirror. Place a pin upright in front of the mirror. Mark its position.

2 Line up one edge of a ruler with the image of the pin in the mirror. Do this again from another position. Mark the position of the ruler edge each time.

3 Take away the mirror, pin and ruler. Find out where the ruler lines would cross. This is the position of the image.

Periscope

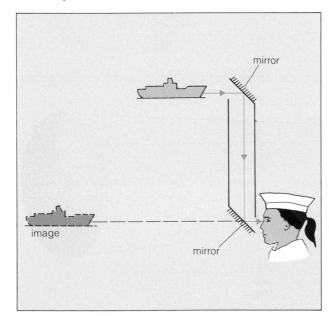

A periscope uses two mirrors to give you a higher view than normal. The image you see is the right way round because one mirror cancels out the lateral inversion of the other.

Note: some periscopes use reflecting prisms rather than mirrors. You can find out more about reflecting prisms on page 117.

1 Copy the diagram.

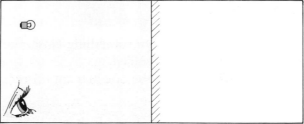

Mark the position of the image of the bulb. Draw two rays which leave the bulb, reflect from the mirror and enter the eye.

2 This is a plan of a room.

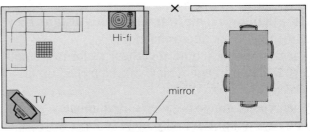

There is a mirror on the wall. If you stood in the doorway at x, would you be able to see:
a the television **b** the Hi-fi?

3 Police cars sometimes have the word STOP on them. It is written in 'mirror writing', so that drivers can read it in their driving mirrors. Write the word STOP as it appears on the police car.

3.3 Curved mirrors

Magic spoons

The same spoon. But very different images.
Try it for yourself. Then put your thumb really close to each side of the spoon and see how the images compare.

Concave mirrors

Concave mirrors curve inwards.
They can form two types of image:

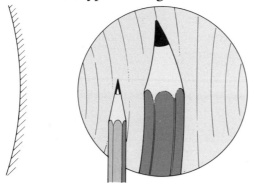

If the light rays come from something close, the image is upright and magnified.
It is a **virtual** image, like the one in a flat mirror.

Convex mirrors

Convex mirrors bulge outwards. They only give one type of image. It is always small, upright and virtual.

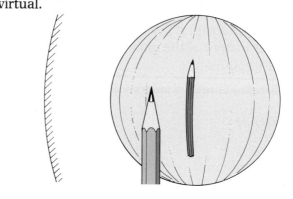

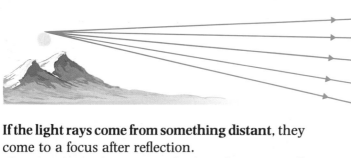

real image

If the light rays come from something distant, they come to a focus after reflection.
They **converge** (come together) to form a small upside-down image which can be picked up on a screen.
This type of image is called a **real** image.
The rays of light actually meet to form it.
Rays from very distant things are nearly parallel to each other. A concave mirror brings parallel rays to a focus at a point called the **principal focus**. The distance from the mirror to the principal focus is called the **focal length**.
Highly curved mirrors have short focal lengths.

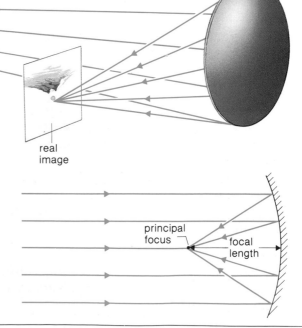

principal focus

focal length

Light isn't the only thing to be reflected by a curved surface. Curved reflectors are also used for sound, radiated heat, and radar and TV signals.

Concave reflectors

A concave mirror is very useful when making up or shaving, because it can magnify.

Not just for decoration. The concave ears are just the job for focusing distant sounds.

1 Which kind of mirror:
a can magnify; **b** always gives a small upright image; **c** can give a real image on a screen?
2 Copy and complete the table to show which type of mirror you would pick for each job.

Use	Type of mirror	Reason for choice
Shop security mirror		
Make-up mirror		
Headlamp reflector		

Radar pulses from a distant aircraft are focused by this concave dish.

Convex reflectors

Convex mirrors give you a much wider view. They are used as driving mirrors and shop security mirrors.

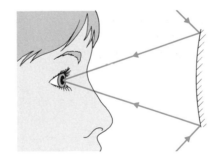

3 This portable gas fire is used by campers.

What type of reflector does it have?
Why does it have a reflector?
Where is the principal focus of the reflector?

3.4 Bending light

It doesn't hurt!

Turn the glass block, and out comes a piece of finger. Or so it seems. In fact, it is the light rays which move, not the finger.

This is how a ray of light passes through a glass block:

The ray is bent, or **refracted**, when it goes into the block. It is bent again when it leaves the block. So, the block moves the ray sideways.

The only time the ray doesn't bend is when it strikes the face of the block 'square on', at right angles.

When light goes into glass, water, or other transparent material, it bends towards the normal. In other words:

the angle of refraction is less than the angle of incidence.

When light leaves a transparent material, it bends away from the normal.

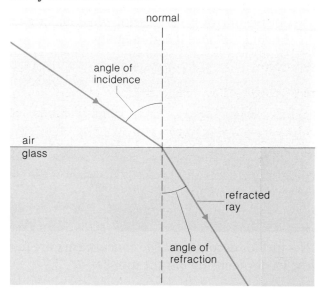

Deeper than it looks

Water never looks as deep as it really is:

Light rays from the pebble bend away from the normal when they leave the water. From above, the rays seem to come from a point which isn't so deep, and is slightly to one side. So, the pebble seems closer than it actually is.

Because of this, objects seem larger underwater. When scientists or archaeologists are working on marine life or wrecks, they overestimate sizes and must measure for accuracy.

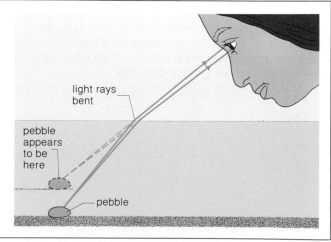

Colour

Where colours come from:

When a narrow beam of white light passes through a prism, the beam splits into all the colours of the rainbow. White isn't a single colour at all, but a mixture of colours. The colours enter the prism together but are bent different amounts by the glass. The effect is called **dispersion**.

The range of colours is called a **spectrum**. Most people think they can see six colours in the spectrum:

red, orange, yellow, green, blue, violet,

though really, there is a continuous change of colour from beginning to end.

The different colours are actually light waves of different wavelengths. Red light has the longest wavelength, blue the shortest.

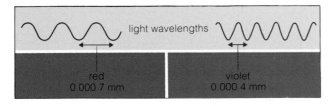

light wavelengths

red
0.000 7 mm

violet
0.000 4 mm

1 Copy the diagrams below. Complete them to show what happens to each ray of light as it passes through the glass block.

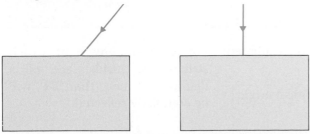

2 A game for the school fête. Scatter some 20p coins over the bottom of a bowl.

Why light bends

A fast car drives at an angle into sand.
One front wheel strikes the sand before the other.
So one side of the car is slowed down before the other.
The path of the car is bent.
Light isn't solid like a car.
But it too bends because it is slowed down.
The more it is slowed, the more it bends.

In air, light travels at about 300 000 kilometres/ second.
This is its speed in some other materials:

Material	Speed of light
Water	225 000 km/s
Glass	197 000 km/s
Perspex	201 000 km/s
Diamond	124 000 km/s

(These speeds vary slightly depending on the colour.)

To play, you take aim from the side, and throw in a 2p coin. You win if your coin covers a 20p coin. Explain why you don't have much chance of winning if the bowl is filled with water.

3 The table on this page gives the speed of light in water, glass, Perspex and diamond.
a Which of these materials bends light the most?
b Compare water with glass. Which of these two bends light more?
c If the blocks in question 1 were made of Perspex instead of glass, how would your drawings be different?
4 Which colour is:
a bent the most by a prism?
b bent the least by a prism?

3.5 Inside reflections

Diamond. The hardest substance known. And probably the most beautiful. It takes a skilled diamond cutter to reveal the true beauty of a diamond. The faces or **facets** have to be cut at carefully chosen angles so that any light going into the diamond is reflected back out again. That is the secret of the sparkle.

Internal reflection

The inside surface of a diamond or a block of glass or water can act like a perfect mirror. It all depends on the angle at which light rays strike it.

Three rays leave a lamp on the bottom of a swimming pool. Each leaves at a different angle:

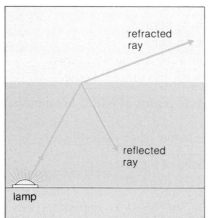 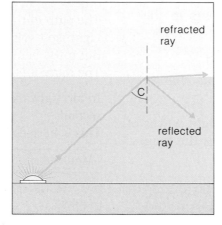 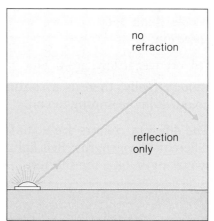

This ray splits.
Some of the light is reflected.
Some is refracted.

This ray splits.
Some of the light is reflected.
Some is refracted. But the angle is so large that the light only just manages to leave the surface of the water.

All of the light is reflected.
There is no refracted ray because the light strikes the surface at too great an angle.

In the middle diagram angle C is called the **critical angle**. If light strikes the surface at a greater angle than this, the surface acts like a perfect mirror. The light is **totally internally reflected**.

Different materials have different critical angles:

Material	Critical angle
Water	49°
Glass	42°
Diamond	24°

For example, any light which strikes the inside face of a diamond at more than 24° will be completely reflected.

Prisms, pipes, and mirages

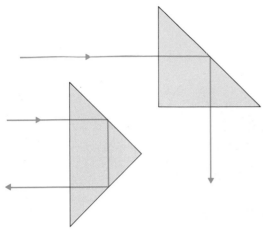

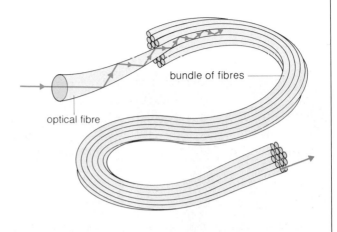

optical fibre

bundle of fibres

Glass prisms can behave like perfect mirrors. But the light has to strike an inside face at an angle greater than the critical angle.

With a light pipe, you can see round corners. Surgeons use them to inspect the insides of stomachs and wombs. Each pipe is a bundle of fine glass or plastic fibres. When a ray enters a fibre at one end, it reflects off the sides until it comes out of the other end. For more on optical fibres, see page 227.

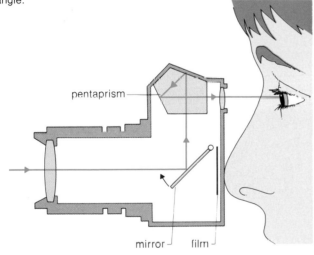

pentaprism

mirror film

With this camera, you actually look out through the lens when you line up your shot. A five-sided 'pentaprism' relects light from the mirror into your eye. When you press the button, the mirror flicks up out of the way so that light can reach the film.

In a hot desert, you may see pools of water in the distance. Only they aren't really there. The image is just a mirage. Just above the ground there is a layer of very hot thin air. This acts like the inside face of a glass block. It reflects light. You see the reflection of the sky, and think that the ground is wet and shiny.

1 Copy and complete each diagram to show what happens to the ray of light after it reaches the prism.

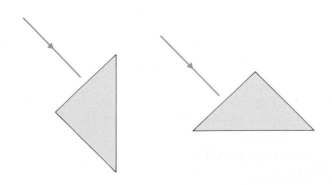

2 A light ray strikes the inside surface of material X.

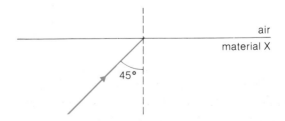

air

material X

45°

Look at the table of critical angles on the opposite page. Then copy and complete the diagram to show what would happen to the ray of light if material X were:

a water **b** glass **c** diamond

3.6 Convex lenses

What do they have in common?

They have a small upside-down image inside them. It's formed by a convex lens.

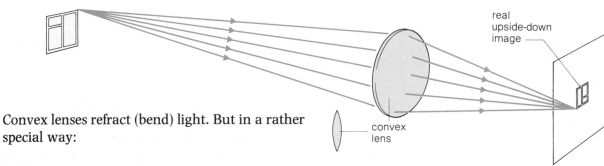

real upside-down image

convex lens

Convex lenses refract (bend) light. But in a rather special way:

Light rays leave a distant window and pass through a convex lens. The rays are bent by the lens. They converge (come together) to form a small upside-down image of the window. The image is **real**. It can be put on a screen.

Rays from very distant things are parallel to each other. A convex lens brings parallel rays to focus at a point called the **principal focus**.
The distance from the lens to a principal focus is called the **focal length**.

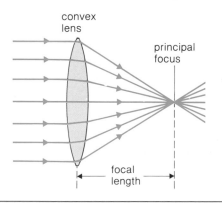

convex lens

principal focus

focal length

How to measure a focal length

- Place a convex lens several metres from a window.
- Focus the image of a distant building on a screen.
- Measure the distance from the lens to the screen. This is the focal length.

Do the experiment twice, with a thick lens and a thinner lens. Which has the longer focal length? Which gives the larger image?

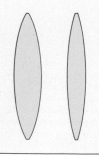

Finding the image

Move something closer to a lens, and its image moves too. You can find the image position by experiment.

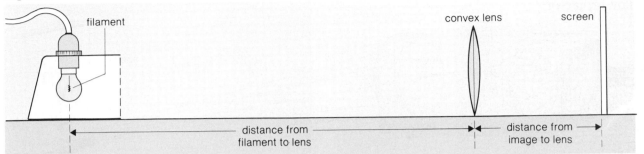

- Place a ray box so that the filament is one metre away from a convex lens.

- Move the screen until you focus the image of the filament.

- Measure:
  the distance from the filament to the lens;
  the distance from image to the lens.
  The length of the image.

- Move the ray box closer to the lens, 10 cm at a time.
  Take measurements as before.
  Enter your readings in a table.

Distance from filament to lens in cm	Distance from image to lens in cm	Length of image in cm
100		
90		
80		
...		

As the filament moves closer to the lens, what happens to the image distance? what happens to the image size?

To answer these questions, you will need to have done the two experiments on this double page.

1 Copy and complete the diagram to show what happens to the rays of light when they pass through the lens.

Mark on the principal focus of the lens and the focal length.

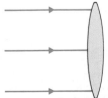

This lens is thicker and more highly curved than the previous one. Copy and complete the diagram.

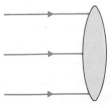

2 Dave did an experiment like the one on this page. Unfortunately he wrote down the image distances on a piece of rough paper. Then he copied them into his book in the wrong order:

Reading	Distance from filament to lens in cm	Distance from image to lens in cm
1	100	24
2	60	60
3	40	30
4	30	20
5	20	18

a Rewrite Dave's table so that the image distances are in the correct order.
b For which reading would you expect the image to be biggest?
c For which reading would you expect the image to be smallest?
d What is the shortest distance between the image and the filament?

3.7 Lenses at work

The projector

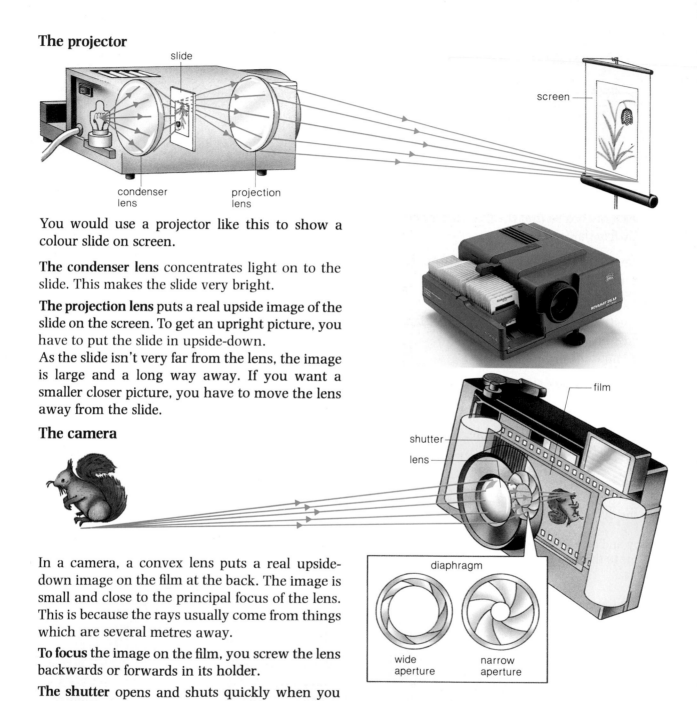

You would use a projector like this to show a colour slide on screen.

The condenser lens concentrates light on to the slide. This makes the slide very bright.

The projection lens puts a real upside image of the slide on the screen. To get an upright picture, you have to put the slide in upside-down.
As the slide isn't very far from the lens, the image is large and a long way away. If you want a smaller closer picture, you have to move the lens away from the slide.

The camera

In a camera, a convex lens puts a real upside-down image on the film at the back. The image is small and close to the principal focus of the lens. This is because the rays usually come from things which are several metres away.

To **focus** the image on the film, you screw the lens backwards or forwards in its holder.

The shutter opens and shuts quickly when you press the button. This lets a small amount of light into the camera.

The film is coated in chemicals which are sensitive to light. The chemicals are changed by the image. Later on, these changes can be 'fixed' to form the photograph.

The diaphragm is a ring of sliding metal plates. It changes the size of the **aperture** (hole) through which the light rays pass. In bright sunshine, a small aperture might be used to cut down the amount of light reaching the film. In many cameras, aperture adjustment is automatic.

The human eye

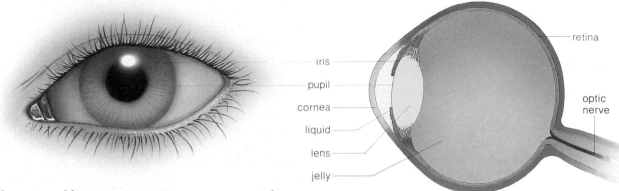

retina

iris
pupil
cornea
liquid
lens
jelly

optic
nerve

The eye is like a camera. It uses a convex lens system to form a small upside-down image of anything in front of it.

The iris controls the amount of light going into the eye. If you walk into a dark room, the hole in the middle of the iris (the **pupil**) grows larger.

The cornea and the watery liquid behind it do most of the focusing of the rays.

The lens itself makes small focusing adjustments. It doesn't move backwards or forwards like the lens in a camera or projector. Instead it becomes thinner or fatter.

The retina is the 'screen' which detects the image. It contains millions of tiny cells which are sensitive to light. The cells send signals along the optic nerve to the brain.

Your brain gives you an upright view of the world. But it isn't always the same as the image in your eye. Look at the examples below.

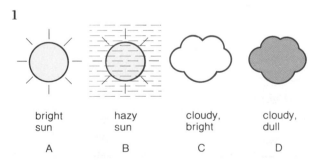

The brown lines look bent. Now check them with a ruler.

The brown lines look the same length. Now check them with a ruler.

1

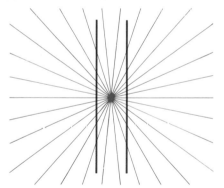

| bright sun | hazy sun | cloudy, bright | cloudy, dull |
| A | B | C | D |

Diagrams A, B, C, and D above represent four different weather conditions.

For which of them would a camera need:
a the widest aperture   **b** the narrowest aperture?
(Assume that the shutter speed does not change.)

2 Copy and complete the chart to show which parts of the camera and the eye do similar jobs:

Camera	Eye	Job done
		focuses rays
diaphragm		
		picks up image

3 Explain why:

a the image from a projector is much bigger than the image in a camera;

b the pupils of your eyes become smaller if you walk out into bright sunshine.

3.8 Waves

Havoc in Alaska

The after-effects of a tsunami or 'tidal wave'. It started with a violent underwater earthquake thousands of kilometres away. This set waves racing across the ocean. Far out at sea, the waves were small. Ships moved up and down a metre or so, but no more. But as the waves approached the coast, they grew to an enormous size. Finally, taller than a house, they hit the shore. As they collapsed, they caused a forward rush of water which carried trees, boulders and boats hundreds of metres inland.

Waves carry energy from one place to another.

They don't only move across water. Sound, light, and radio signals all travel in the form of waves.

There are two main types of wave. You can study them using a stretched 'Slinky' spring.

Transverse waves

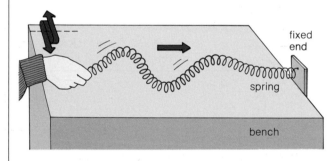

Keep moving one end of the spring from side to side, and waves travel along the spring. Each coil moves from side to side but a little later than the one before. Waves like this, where the movements are sideways (or up and down) are called transverse waves.

Longitudinal waves

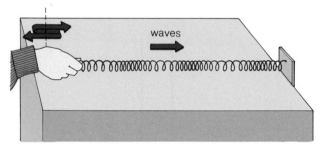

Keep moving one end of the spring backwards and forwards, and waves travel along the spring. Each wave is a compression followed by a stretched-out section. Waves like this, where the movements are backwards and forwards, are called longitudinal waves.

How to draw waves

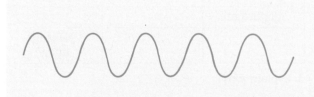

Transverse waves can be drawn like this.

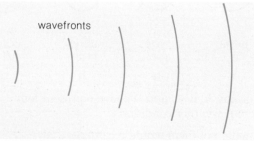

You can also draw waves using lines called wavefronts. Think of each wavefront as the top of a transverse wave, or the compression of a longitudinal wave.

Describing waves

The **wavelength** is the distance between wave-fronts ... or between any place on one wave and the same place on the next.

The number of waves passing every second is called the **frequency**. It is measured in **hertz (Hz)**. A frequency of 100 Hz means that 100 waves are passing every second.

This height shown on the right is called the **amplitude** of the wave.

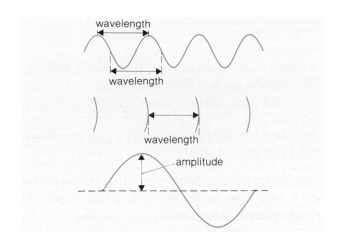

An equation for waves

Imagine waves travelling across the sea ...

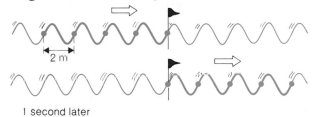

I second later

speed = frequency × wavelength

This equation is true for all waves.

Here, 4 waves pass the flag in one second ...
so the frequency is 4 Hz.
Each wave is 2 metres long ...
so the wavelength is 2 m.

This means that:
The waves travel 8 metres in one second ...
so the speed is 8 m/s.

In this example,

8	=	4	×	2
m/s		Hz		m

1

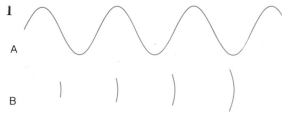

a What type of wave is A?
b Using a ruler marked in millimetres, measure:
the wavelength of A;
the amplitude of A;
the wavelength of B.

2 Three waves travel at the same speed, but they have different frequencies and wavelengths. Copy the chart, then fill in the blank spaces:

	Speed in m/s	Frequency in Hz	Wavelength in m
Wave 1		8	4
Wave 2		16	
Wave 3			1

3 These waves all travel at the same speed:

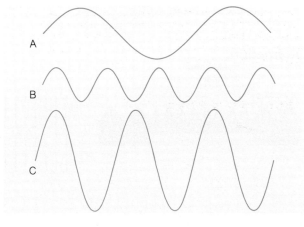

Which has:
a the highest frequency?
b the longest wavelength?
c the greatest amplitude?

3.9 Ripples of water and light

A **ripple tank** – for studying how waves behave. The shallow tank is filled with water.

The vibrating dipper sends ripples across its surface.

You place different shapes in the water to reflect or bend the wave 'beam'.

The ripples seem to behave in much the same way as a beam of light. It's one good reason for thinking that light is made up of waves.

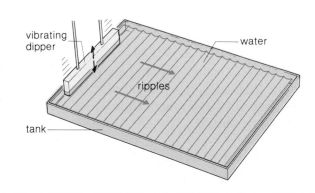

In water

In light

Reflection

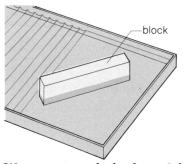

Waves approach the face of the block.

The angle of reflection is the same as the angle of incidence – as with a beam of light.

A beam of light is reflected by a mirror.

Refraction

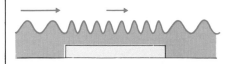

A flat piece of plastic makes the water less deep. This slows down the waves. As they slow down, they bend – just like a

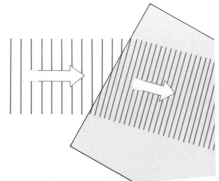

beam of light. Use a lens-shaped piece of plastic, and the waves are focused.

A beam of light bends when it enters glass.

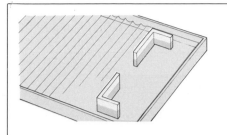

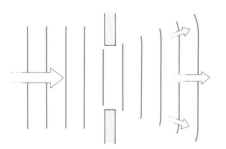

Diffraction

Waves bend when they pass through a narrow gap. This is called **diffraction**. It works best if the width of the gap is about the same as the wavelength. Wide gaps don't cause much diffraction.

Try looking at a street light through the cover of an umbrella. You will see many images. These are caused by diffraction. The light is bent off course as it passes through the tiny holes in the material.

Gaps have to be extremely small to diffract light. What does this tell you about the wavelength of light waves?

Interference

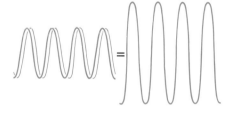

waves add

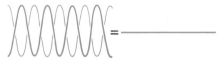

waves cancel

Surfers know the problem well. Sometimes sea waves combine together to produce a very large wave. Sometimes, they cancel each other out. This is called interference. To see interference in a ripple tank, you send out ripples from two points. You get an **interference pattern**. In some places, the ripples add together; in others they cancel out.

Try looking hard at a soap bubble just before it pops. You should see dark patches spread across it. This is because of interference. Light waves reflected from the outside of the bubble are cancelled by waves reflected from the inside surface.

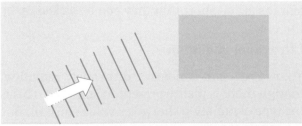

1 Waves in a ripple tank approach a rectangular block of plastic. The water covers the plastic.
a What happens to the speed of the waves when they reach the plastic? What happens to the waves?
b Copy and complete the diagram to show what happens to the waves.
c The plastic is replaced by another block which is deeper than the water. Draw a diagram to show what now happens to the waves.

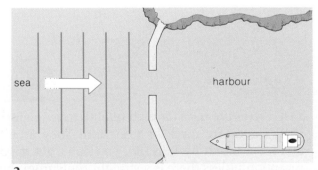

sea   harbour

2
a Copy and complete the diagram to show what happens to the waves when they pass through the harbour entrance.
b What is this called?
c What difference would there be if the harbour entrance were wider?

3.10 Electromagnetic waves

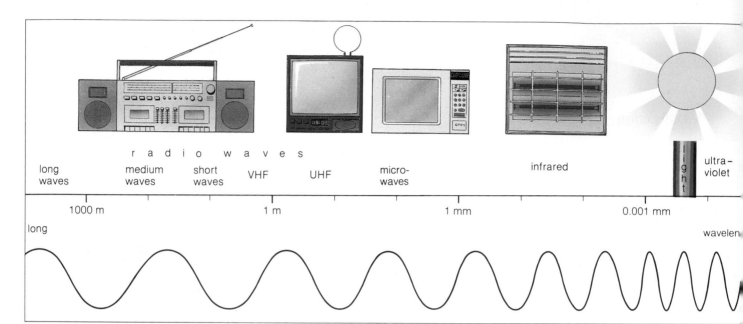

| long waves | medium waves | short waves | VHF | UHF | micro-waves | infrared | light | ultra-violet |

radio waves

1000 m · 1 m · 1 mm · 0.001 mm

long · wavelength

Radio waves

Radio waves are produced by making electrons vibrate in an aerial. They can't be seen or heard. But they can be sent out in a special pattern which tells a TV or radio what pictures or sounds to make.

Long and medium waves will diffract round hills. Your radio will pick them up, even down in a valley.

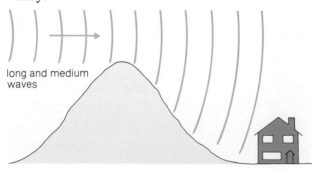

long and medium waves

VHF waves are used for high-quality stereo radio; **UHF waves** are used for television. They aren't diffracted much by hills. So, you don't get good reception unless there is a straight route from the transmitter to your radio or TV aerial.

Microwaves are very short wavelength radio waves. They are used for radar, and for sending signals to satellites. They are also used for beam-ing television and telephone signals round the country.

Some microwaves are absorbed strongly by food. This produces heat. The idea is used in microwave ovens.

Infrared radiation

Hot things like fires and radiators all give off infrared radiation. In fact, everything gives off some infrared. Usually it comes from molecules which are vibrating rapidly.

As molecules get hotter, infrared wavelengths get shorter. When something is 'red hot', some wavelengths are so short that they can be picked up by the eye.

Ultraviolet radiation

Your eyes can't detect ultraviolet radiation, though there is plenty present in sunlight. This is the type of radiation that gives you a sun-tan. But too much can damage your eyes and your skin.

Some chemicals glow when they absorb ultraviolet. The effect is called **fluorescence**. This is the secret of 'whiter than white' washing powders. They absorb the ultraviolet in sunlight. Then the chemicals glow to make your clothes look brighter than normal.

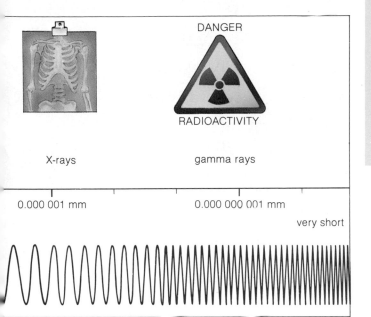

X-rays   DANGER   gamma rays

RADIOACTIVITY

0.000 001 mm   0.000 000 001 mm

very short

X-rays

X-rays can be produced using an X-ray tube. The X-rays come out when fast-moving electrons smash into a metal target.

Short wavelength X-rays are extremely penetrating. They can even pass through dense metals like lead.

Long wavelength X-rays are less penetrating. They can pass through flesh but not bone. So, bones will show up on an X-ray photograph.

All X-rays are dangerous because they can damage living cells deep in the body.

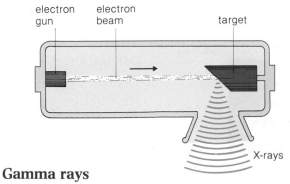

electron gun   electron beam   target

X-rays

Gamma rays

These are given off by radioactive materials. Gamma rays have the same effects as X-rays and are extremely dangerous.

Light is a member of a whole family of waves called the **electromagnetic spectrum**. These waves have several things in common:

- they can travel through empty space;
- they travel through space at the same speed; 300 000 km/s
- they are electric and magnetic ripples, mostly given off by electrons or molecules as they vibrate or lose energy. (Electrons are tiny electrical particles that come from inside atoms – see page 150.)

1 When the beam from the filament passes through the glass prism, two other types of radiation can be detected, as well as light.
Which type of radiation is at X?
Which type of radiation is at Y?

2 Name a type of electromagnetic wave which:
a can cause fluorescence;
b is diffracted by hills;
c is used for radar;
d can pass through metals;
e is given off by hot materials;
f can be detected by the eye.

3

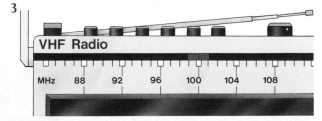

VHF Radio

MHz   88   92   96   100   104   108

Speed of electromagnetic waves = 300 000 000 m/s
1 MHz = 1 000 000 Hz
Speed = frequency × wavelength

Use the information above to calculate the wavelength of the waves being picked up by the radio.

3.11 Heat radiation

Taking it in ...

Absorbing electromagnetic waves.
The waves come from the Sun. They are mostly infrared, light and ultraviolet. And they warm up anything (or anyone) that absorbs them. They're known as 'heat radiation' – or just 'radiation' for short.

Some surfaces are better at absorbing radiation than others:

Standing in the sunshine, a black car warms up more quickly than any other. Touch the bodywork to test it for yourself.

... sending it out

Radiating electromagnetic waves in London.
Hot marathon runners give off radiation after the race. The more they radiate, the more body heat they lose. And that means a risk of chilling. Silvery sheets solve the problem. They're a quick and easy way of keeping everybody warm.

Some surfaces are better at sending out or **emitting** radiation than others:

A black saucepan cools down more quickly than any other. Could you design an experiment to test it for yourself?

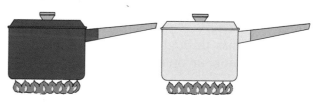

Good absorbers of heat radiation are also good emitters. This is how different surfaces compare:

Dull black surfaces are the best absorbers of radiation. They reflect hardly any radiation at all.

Shiny silvery surfaces are the worst absorbers of radiation. They reflect nearly all the radiation that strikes them.

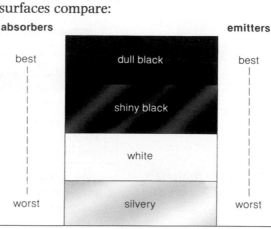

absorbers		emitters
best	dull black	best
	shiny black	
	white	
worst	silvery	worst

Dull black surfaces are the best emitters of radiation.

Silvery surfaces are the worst emitters of radiation.

Keeping your food warm

Shiny aluminium foil helps keep food dishes warm when they're out on the table.

Keeping your plants warm

How a greenhouse traps heat:

Short wavelength radiation from the Sun passes easily through the glass of a greenhouse. It warms the plants inside. The warm plants also radiate heat, but the wavelengths are longer and don't pass so easily out through the glass.
Result: a build-up of heat in the greenhouse.

1

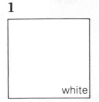
white

dull black

shiny black

Which of these surfaces is best at:
a absorbing heat radiation;
b emitting heat radiation;
c reflecting heat radiation.

2 All three kettles are the same except for their outside surfaces. All are full of boiling water.

dull black

silvery

shiny black

Keeping your drinks warm

outer glass skin

inner glass skin

gap: air removed

silvery surfaces

Shiny kettles keep their heat longer than other sorts.

This Thermos flask has two glass 'skins' with a gap between them. Air has been removed from the gap to stop heat escaping by conduction. The silvery surface stops heat escaping by radiation.

> Heat can travel in three ways:
> by conduction, by convection, and by radiation.
> Read about the other two on pages 90 to 93.

When the kettles are left to cool, this is what happens to their temperature:

	Temperature in °C		
	Kettle A	Kettle B	Kettle C
	100	100	100
After 5 min	85	90	80
After 10 min	73	82	65

a Which kettle has the silvery surface?
b Which kettle has the dull black surface?
3 Explain why:
a in hot countries, houses are often painted white;
b on a hot summer's day, the inside of a white car is cooler than a dark one;
c if you use a lens to focus the Sun's rays on newspaper, the print burns more easily than the white paper.

3.12 Seeing colours

The retina of a human eye contains millions of light-sensitive cells. These are called **rods** and **cones** because of their shapes. There are more rods than cones. Rods respond to weak light, but cannot detect colour. Cones are concentrated in the centre of the retina. They need more light to start working, but are sensitive to colour.

Humans can see hundreds of colours. Yet the retina has only *three* types of colour-sensitive cell:

- Cones which are switched on by **red** light.
- Cones which are switched on by **green** light.
- Cones which are switched on by **blue** light.

Each type is also sensitive to other colours in their part of the spectrum. So between them, they cover the full range. The brain senses *all* colours by how they affect each type of cone. So you see all colours in terms of **red**, **green**, and **blue**. These are the **primary** colours.

Adding colours

On the right, beams of red, green, and blue light are overlapping on a white screen.

White   This is seen where all three primaries overlap. So:

red + green + blue = white

Pure white contains all the colours of the rainbow. But a mixture of red, green, and blue is enough to give the eye the sensation of white. It switches on all three types of colour-sensitive cell, just like pure white would.

Secondary colours   Where any two primary colours overlap, a new colour is seen:

red + green = yellow
green + blue = cyan
red + blue = magenta

The colours **yellow**, **cyan**, and **magenta** are called the **secondary** colours. They aren't really single colours. They just seem that way to the brain. (However, the yellow seen in the spectrum is not a mixture of other colours.)

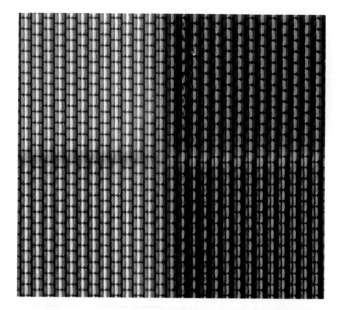

By adding red, green, and blue light in different proportions, the brain can be given the sensation of almost any colour. The effect is used in colour television. The screen of a colour TV (like the one above) is covered with thousands of tiny red, green, and blue strips. These glow in different combinations to produce a full-colour picture.

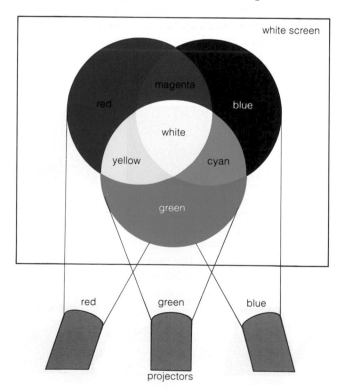

Taking colours away

Most things do not give out their own light. We see them because they reflect light from the Sun (or some other source). To the eye, white sunlight is a mixture of red, green, and blue. Things look coloured in sunlight because they *reflect* only some colours and *absorb* the rest. In other words, they take colours away from white.

In white light:

- A *red* cloth reflects red light but absorbs green and blue.

- A *yellow* cloth reflects red and green light but absorbs blue.

- A *white* cloth reflects red, green, and blue. It absorbs no colours.

- A *black* cloth reflects virtually no light. It absorbs red, green, and blue.

Paints reflect some colours and absorb others. When paints are mixed together, more colours are absorbed, rather than added. So the final colour is not the same as you get when light beams overlap. That is why artists have a different set of colours which they call their 'primaries'.

Filters are pieces of plastic or glass which let through certain colours only. The red filter below *transmits* red light but *absorbs* green and blue.

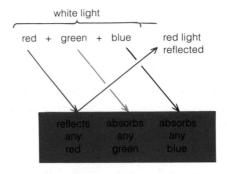

In white light, a red cloth looks red

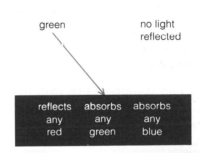

In green light, the same cloth looks black

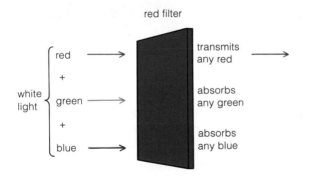

If you shine coloured light on something, its colour may change. There is an example of this above. A cloth which looks red in white light, looks black when viewed in green light. If green light only is striking it, then that is absorbed, so no light is reflected.

1 What are the three primary colours?
2 What colour is produced when the three primaries are added together?
3 What colour is produced when red and green light beams overlap on a white screen?
4 **What colours does a white coat reflect?**

5 What colours does a black coat absorb?
6 What colour does a blue filter transmit? What colours does it absorb?
7 What colour does a green coat reflect? What colours does it absorb?
8 What colour will a green coat appear in red light?

3.13 Light and sight

Bigger ...

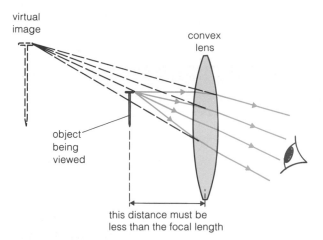

virtual image

convex lens

object being viewed

this distance must be less than the focal length

A convex lens forms a magnified, upright, and virtual image of anything put close to it. This idea is used in a **magnifying glass** as shown above. The rays aren't brought to a focus. They diverge. But, to the viewer, they appear to come from a position behind the lens. (Compare this with the situation shown on page 118, where a convex lens brings distant rays to a focus.)

Darker ...

Sunglasses reduce the amount of light entering your eyes. Some are made from Polaroid, which reduces glare. You can find out how on the right.

Poor-quality sunglasses can be harmful because they let through the invisible ultraviolet in sunlight. When you wear sunglasses, your pupils become larger because of the darker conditions. So, if the ultraviolet is not blocked, more of this damaging radiation is able to reach the retina than normal.

Polarization

Like ripples on water, light waves are transverse waves. However, most light is a mixture of waves which vibrate up and down, side to side, and all planes in between. Light like this is called **unpolarized** light.

Polaroid is a material which blocks wave vibrations in all planes except one. When light leaves a piece of Polaroid, its wave vibrations are in one plane only (for example, up and down). Light like this is called **polarized** light.

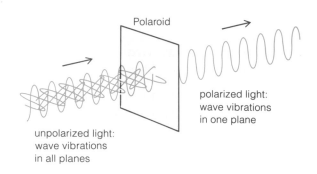

Polaroid

polarized light: wave vibrations in one plane

unpolarized light: wave vibrations in all planes

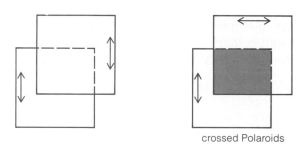

crossed Polaroids

If you put one piece of Polaroid over another, and the planes of vibration match, the light can pass through and the Polaroids look transparent (clear). But if you rotate one Polaroid through 90°, polarized light from the first cannot pass through the second. Arranged like this, the Polaroids are **crossed**. They block all light, and look black.

When light is reflected from water or glass it becomes partly polarized. Polaroid sunglasses block some of this reflected light, which is why they reduce glare from wet surfaces.

The 'liquid crystal' displays on calculators use polarized light. The liquid is sandwiched between two strips of Polaroid. The strips are crossed, but they look clear because the liquid rotates the wave vibrations through 90°. However, if a tiny current is passed through the liquid, the rotation effect is stopped, and the crossed Polaroids look black. By sending currents through different areas of the liquid, black patches can be produced. These form the different numbers of the display.

Short and long sight

Some people's eyes cannot produce a clearly-focused image on the retina. Spectacles (or contact lenses) are needed to solve the problem.

Short sight In a short-sighted eye, the eye lens cannot become thin enough for looking at distant things. So the rays are bent inwards too much. They meet before they reach the retina.

Long sight In a long-sighted eye, the eye lens cannot become thick enough for looking at close things. So the rays are not bent inwards enough. On reaching the retina, they have still not met.

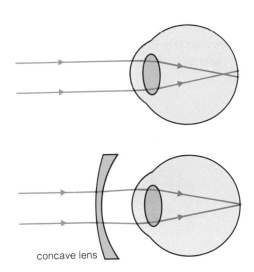

concave lens

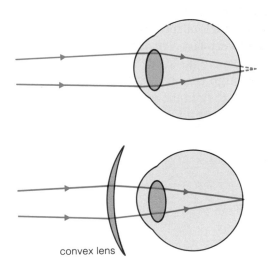

convex lens

A *concave* spectacle lens solves the problem. It bends the rays outwards a little before they enter the eye.

A *convex* spectacle lens solves the problem. It bends the rays inwards a little before they enter the eye.

1 What type of lens would you use for a magnifying glass?
2 What type of image is formed by a magnifying glass? How is it different from the image formed by the lens when the rays entering it come from something a long way away?
3 What is the difference between polarized light and unpolarized light?

4 Why do Polaroid sunglasses reduce the glare from the surface of a puddle?
5 Give an example of the use of polarized light.
6 What is the difference between a short-sighted eye and a normal eye.
7 What type of spectacle lens does a short-sighted person need? Why?

3.14 Sound waves

When these vibrate . . .

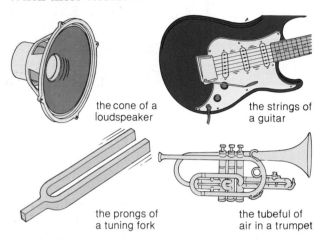

the cone of a loudspeaker

the strings of a guitar

the prongs of a tuning fork

the tubeful of air in a trumpet

. . . they give off sound waves.

Sound waves are vibrations.
When a loudspeaker cone vibrates it moves in and out very fast. This stretches and squashes the air in front. The 'stretches' and 'squashes' travel out through the air as waves. When they reach your ears, they make your ear-drums vibrate and you hear a sound.

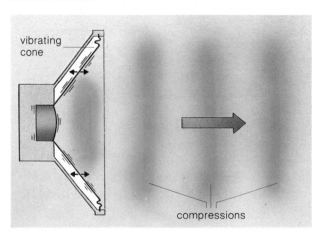

vibrating cone

compressions

Sound waves are longitudinal waves; the vibrations are backwards and forwards (sée page 122).

The 'squashes' are called **compressions**; the air pressure here is higher than normal.

The 'stretches' are called **rarefactions**; the air pressure here is lower than normal.

Sound waves can travel through solids.
They can travel through doors, floors, ceilings and brick walls.

Sound waves can travel through liquids. You can still hear sound when you are swimming underwater.

Sound waves can travel through all gases. This flask has air in it. But, you would still hear the bell ringing whatever type of gas was in the flask.

Sound waves can't travel through a vacuum (empty space). If the air is pumped out of the flask, the sound stops, even though the bell goes on working. Sound waves can't be made if there is nothing to be squashed and stretched.

Seeing sounds

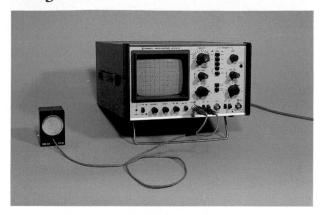

You can't see sounds. But with a microphone and an oscilloscope, you can show sounds as wave shapes on a screen.

When sound waves reach the microphone, they make a tiny sheet of metal vibrate. The microphone changes the vibrations into electrical vibrations. The oscilloscope uses these to make a spot vibrate up and down on the screen. It moves the spot steadily sideways at the same time. The result is a **waveform**.

The waveform looks like a series of transverse waves. But it is really a graph of pressure against time. It shows how the air pressure near the microphone rises and falls as sound waves pass.

If the sound gets louder, the amplitude of the waveform increases.

1 Someone blows a whistle near a microphone. This is the waveform produced on the screen of an oscilloscope.

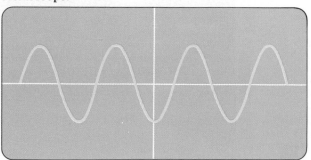

Recorded sounds

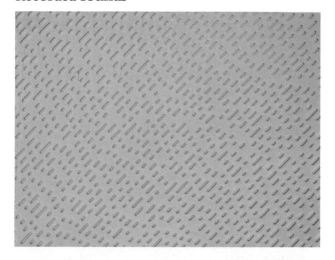

The playing surface of a **compact disc (CD)** is covered with millions of tiny steps, arranged in a track which spirals outwards from the centre. When the disc rotates, a tiny beam of laser light is reflected from the steps and the gaps between. As a result, a series of light pulses is picked up by a detector and changed into electrical signals for processing. The processed signals represent numbers. If you plotted them, the graph would be the waveform of the recorded sound. A CD player doesn't plot graphs! But it does process the number signals to give electrical signals which will make a loudspeaker cone vibrate.

Recording using numbers is called **digital** recording. For more on digital signals, see page 226.

a Use a ruler marked in millimetres to measure the amplitude of the waveform.
b Redraw the waveform so that it has an amplitude of 15 mm. Would the whistle that produced this be louder than the one before?
2 Copy the sentences and fill in the blanks:
a Sounds are caused by __.
b Sound waves can't travel through a __.
c Sound waves are __ waves.
d When sound waves pass, the __ of the air rises and falls rapidly.
3 Using things you might find around the house, how could you show someone that sound can pass through solid materials?

3.15 Vibrations

Vibrations don't only make sounds. They can have other effects as well:

Old brickwork is easily damaged by the vibrations from heavy traffic.

Here, damaging vibrations are being put to work – to crack concrete.

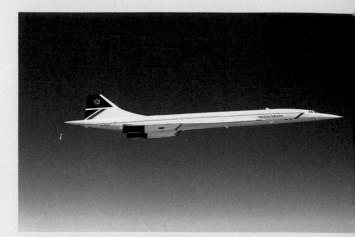

Supersonic aircraft aren't allowed to fly fast over populated areas. When planes fly faster than sound they cause air vibrations which can shatter windows.

Road vibrations can make a journey uncomfortable and tiring. But not for this truck driver. That's because the driving seat is suspended on a cushion of air.

Resonance

If a car wheel is out of balance, it will vibrate when the car reaches a certain speed. This 'wheel wobble' makes the steering wheel judder and can damage the suspension. To prevent it, each car wheel has to be balanced by fixing small lead masses to the rim.

Above, and above right, you can see two examples of an effect called **resonance**:

There is some outside source of vibration. And there is something (a car wheel for example) which naturally vibrates at a certain frequency. If the frequency of the outside source matches this natural frequency, then the amplitude of the vibrations builds up to a peak.

Below, you can see another example of resonance.

In washing machines, 'drum wobble' can be a problem. It happens when the clothes aren't spread evenly inside the drum. When the drum reaches a certain speed, severe vibrations start. These may damage the machine or make its moving parts wear out more quickly.

1 If a bus engine is ticking over, the side windows of the bus may start to vibrate strongly.
a What is the effect called? **b** Why are the vibrations less strong when the engine revs up?
2 In each of the photographs below, what is vibrating, and what is the effect used for?

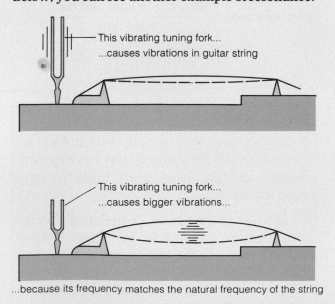

This vibrating tuning fork...
...causes vibrations in guitar string

This vibrating tuning fork...
...causes bigger vibrations...

...because its frequency matches the natural frequency of the string

3.16 Speed of sound

Lightning strikes. You see the flash. The crash comes later. Much later if you're lucky.

Sound is very much slower than light. So you always hear things after you've seen them. Over short distances, you don't notice the difference. But with distant lightning, there can be a delay of several seconds. The longer the delay, the further away the lightning – it's about 3 seconds for every kilometre (half mile).

In air, the speed of sound is about 330 metres/second. That's about four times faster than a racing car, but slower than Concorde.

The speed of sound depends on the temperature of the air. Sound waves travel faster through hot air than through cold air.

The speed of sound doesn't depend on the pressure of the air. If the pressure rises the speed of the waves stays the same.

The speed of sound is different through different materials. Sound waves travel faster through liquids than through gases. They travel fastest of all through solids.

speed of sound

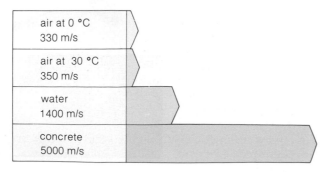

| air at 0 °C 330 m/s |
| air at 30 °C 350 m/s |
| water 1400 m/s |
| concrete 5000 m/s |

Echoes

Hard surfaces like walls reflect sound waves. When you hear an **echo**, you are hearing a reflected sound a short time after the original sound.

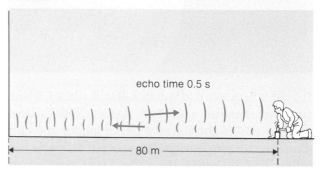

echo time 0.5 s

80 m

This girl is 80 metres from a large brick wall. She is hammering a block of wood. Every time she hits the block, she hears an echo 0.5 seconds later. This is the **echo time**.

She could use this information to calculate the speed of sound:

$$\text{speed} = \frac{\text{distance travelled}}{\text{time taken}}$$

so

$$\text{speed of sound} = \frac{\text{distance to wall and back}}{\text{echo time}}$$

$$= \frac{2 \times 80}{0.5} = \frac{160}{0.5} = 320 \, \text{m/s}$$

Do-it-yourself

If there's a large wall around, you can find the speed of sound for yourself. Just fit your own distance and time measurements into the equation above.

To make your time measurement more accurate, measure the time for 20 echoes instead of just one. Bang the block repeatedly so that each blow is made just as an echo returns. If it takes 10 seconds to make 20 hammer blows, then the echo time is $10 \div 20$ seconds, or 0.5 seconds.

Using echoes

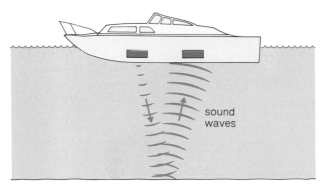

sound waves

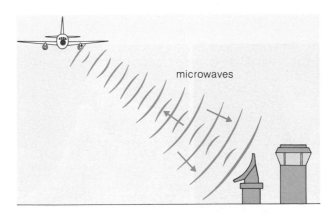

microwaves

Ships use **echo sounders** to measure the depth of water underneath them. An echo sounder sends bursts of sound waves towards the sea-bed. Then it measures the time taken for the echoes to return. The longer the time, the deeper the water.
For example:

If   – a burst of sound takes 0.1 seconds to reach the sea-bed and return,
and – the speed of sound in water is 1400 m/s,
then – distance travelled = speed × time
$$= 1400 \times 0.1 \text{ m}$$
$$= 140 \text{ m}$$
But – the sound has to travel down *and* back,
so   – the depth of water is 70 m.

		distance travelled in m	time taken in s
	rocket	900	3.0
	aircraft	1000	2.0
	bullet	100	0.5
	meteorite	3000	0.1

1 If the speed of sound in air is 330 m/s, which of the above are travelling faster than sound?
2 If the speed of sound in air is 330 m/s, how far does sound travel in:
a 1 second
b 2 seconds
c 10 seconds
d 0.1 seconds?

Radar works rather like an echo sounder. Except that microwaves are sent out rather than sound waves. The microwaves are reflected by aircraft. The longer they take to return, the further away the aircraft.

Losing echoes

Echoes can be a nuisance. In empty rooms, cinemas and concert halls, reflected sounds can take so long to die away that it is sometimes difficult to hear anything clearly. Carpets, curtains and soft chairs help to solve the problem. Modern concert halls are designed so that sounds are neither muffled nor echoing around.

3 Jeff thinks that his cassette player sounds clearer in the bedroom than it does in the kitchen. Is he imagining things? Or could he be right? Explain.
4 The echo sounder in this ship sends a burst of sound waves towards the sea-bed. 0.2 seconds later, reflected waves are picked up by the ship.
a How long did it take the waves to reach the sea-bed?
b If the speed of sound in water is 1400 m/s, how far is it to the sea-bed?

3.17 Sounds high and low

She doesn't necessarily sing better than him.
But she can reach higher notes.
She can give out more sound waves every second.

Frequency and pitch

The number of waves per second is called the
frequency. It is measured in **hertz (Hz)**.
If a singer gives out 200 sound waves every
second, the frequency is 200 Hz.

Different frequencies sound different to the ear.
You hear *high* frequencies as *high* notes.
They have a **high pitch**.
You hear *low* frequencies as *low* notes.
They have a **low pitch**.

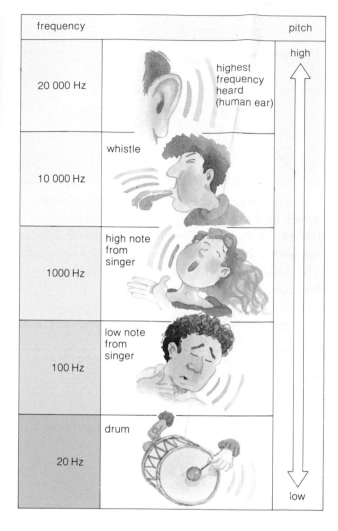

frequency		pitch
		high ↑
20 000 Hz		highest frequency heard (human ear)
10 000 Hz	whistle	
1000 Hz	high note from singer	
100 Hz	low note from singer	
20 Hz	drum	
		low ↓

Changing note

The notes from many musical instruments are all
based on **octaves**.
Each time the pitch goes up an octave, the
frequency doubles.

Notes on a keyboard. Each C is double the frequency of the one to its left. This is also true for
any other notes that are an octave apart.

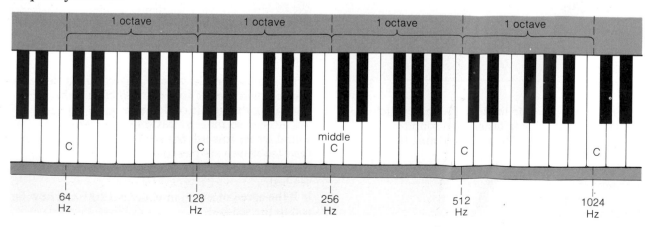

| 1 octave | 1 octave | 1 octave | 1 octave |

C — 64 Hz
C — 128 Hz
middle C — 256 Hz
C — 512 Hz
C — 1024 Hz

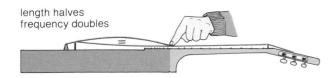

length halves
frequency doubles

A vibrating guitar string.

To make the pitch of the note higher, you could
– tighten the string
– shorten the length of string which can vibrate by
  pressing the string against a fret.

If you halve the vibrating length of the string, the
frequency doubles and pitch goes up by an octave.

The note from this clarinet comes from a vibrating
column of air inside. As you uncover the air holes,
the vibrating column gets shorter and the pitch
goes up.

Quality

Middle C on a guitar doesn't sound quite the same
as Middle C on a piano – and its waveform looks
different on an oscilloscope screen. The two
sounds have a different **quality**.

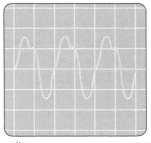
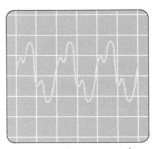

guitar **waveforms** piano

In fact, each sound has a strong **fundamental
frequency**, giving Middle C. But other weaker
frequencies are mixed in as well. These are called
overtones. They differ from one instrument to
another. With a synthesizer, you can choose
which frequencies you mix together, and make
the sound of a piano, guitar or any other instrument you want.

---

speed = frequency × wavelength

speed of sound in air = 330 m/s

1000 hertz = 1 kilohertz (kHz)

1 Copy the chart below. Fill in the blanks to show
the frequencies and wavelengths of the different
sounds:

frequency in Hz	wavelength in m
?	10
?	2
330	?
660	?

2 What happens to the pitch of a guitar string if you:
a shorten the string;
b slacken the string?

3 Four instruments are giving out four different
notes:

instrument	flute	guitar	trumpet	keyboard
frequency	400 Hz	150 Hz	500 Hz	200 Hz

a Which has the highest pitch?
b Which two notes are one octave apart?
c A saxophone plays a note 2 octaves higher than
the guitar. What is the frequency of the note?
d Explain why the saxophone doesn't sound like the
guitar even if both play the same note.

3.18 Noise and ultrasound

Living with noise

The people who live here don't need any scientist to tell them what noise is. It is any sound they don't like. Especially the sound of jet airliners taking off over their garden. Since the runway was extended, the value of their house has dropped. And they have had to fit double-glazing to cut down the noise. They are very dissatisfied, and keep writing to the airport manager to say so.

The Airport Manager thinks that they are being unreasonable. Pilots have to follow special rules during take-off to limit engine noise. And at night, there's a complete ban on flights altogether. Besides which, jets are much quieter than they used to be. Modern jets push out air more smoothly than old ones. So they produce less noise.

What noise? Debbie is protected from noise by the Safety at Work Act. When she is working in the machine shop, her company has to provide her with ear protectors.

Noise damage Strictly speaking it isn't noise at all. Just high-quality rock music. But if you keep the volume turned up for hours on end, it can damage your ears. In extreme cases, it can lead to deafness.

Noise level

Scientists check noise levels using meters marked in **decibels (dB)**. Some typical readings are given in the table below:

	Noise level in dB
Personal stereo, played loud	150
Damage to ears	140
Rock concert	110
Some ear discomfort	90
Telephone ringing	70
Normal speech	60
Whispering	40

Ultrasound

Sounds which are too high for the human ear to hear are called **ultrasonic sounds**, or **ultrasound**. Sounds like this have frequencies above 20 kHz (20 000 Hz).

Below, ultrasound is being used to check an unborn baby in the womb. An ultrasound transmitter is moved over the mother's body. A detector picks up sound waves reflected from different layers inside the body. The signals are processed by a computer, which puts an image on a screen. The method is safer than using X-rays, because X-rays damage body cells.

You want to play your stereo loud. Your next-door neighbour doesn't like your type of music. Write down the ways in which you could reduce the amount of noise reaching your neighbour.

Most cinemas have several films on show in the same building. Try to find out how the noise from one studio is prevented from reaching another.

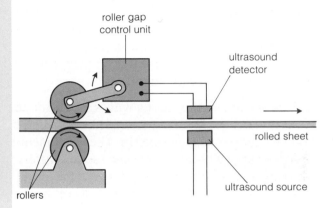

In production processes, it is important to check the things being made to make sure that their size and quality are up to standard. This is called **quality control**. Above, you can see one method of automatic quality control, used in the production of rolled steel. An ultrasound detector controls the gap between the two rollers. Its job is to make sure that the steel is not rolled too thickly or too thinly.

If the steel is rolled too thickly, how will this affect the ultrasound received by the detector?
What will the signals from the detector make the control unit do?
What will happen if the steel is rolled too thinly?

At higher powers, ultrasound is used in hospitals as a treatment for kidney stones — these are rather like the scale in a kettle, but they form in people's kidneys. By directing ultrasound at the kidneys, the stones can be broken up without the patient needing an operation.

Ultrasound is also used in industry. Like X-rays, it will show up cracks in metals. Some industrial cleaning processes make use of ultrasound, as the high-frequency sound waves will dislodge dirt.

3.19 Seismic waves

Earthquakes happen when large natural forces affect rocks so that they move. A break in a rock is called a **fault**. The breaking rocks release energy into the earth, which shakes.

In 1990, there was a small earthquake in Shropshire, England. It was felt by people in many parts of the British Isles. Much stronger earthquakes happen around the world, and although they may not be felt by people in Britain, they can be detected by special equipment.

Seismic waves are the energy waves sent through rocks by earthquakes. They spread like waves on a pond when a stone is dropped in. The instruments that detect the vibrations are called **seismometers**. They are linked to recording systems which 'draw' waveforms called **seismograms**, usually on paper.

Across the world, there is an international network of recording stations like the one above for monitoring seismic waves.

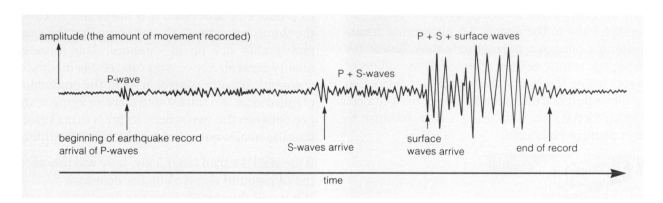

Above, you can see a typical seismogram record from an earthquake. The waves must have started together at the **focus** (site) of the earthquake. But they do not reach the seismometer together because they travel at different speeds.

The first arrivals are **primary waves (P-waves)**. They are *longitudinal* waves (see page 122). Think of the 'P' as standing for **P**ush-pull waves.

Next come **secondary waves (S-waves)**. They are *transverse* waves (see page 122). Think of the 'S' as standing for **S**haking waves. Unlike P-waves, S-waves cannot travel through molten (liquid) rock.

The left-hand diagrams at the top of the next page show how P-waves and S-waves travel.

P-waves and S-waves travel *through* the Earth, but the final group of waves travel the long way, around the curved surface of the Earth. They are the **surface waves**.

The right-hand diagrams at the top of the page show two types of surface wave, named after the two scientists who investigated them.

Surface waves are slower than P-waves and S-waves. But in the earthquake zone, they are the most destructive. Some produce a rolling motion, rather like waves at sea. Huge fissures (cracks) can appear on the crests, and close seconds later when the crests become troughs. People, cars, and even buildings may be swallowed up.

direction of travel of waves ⟶   direction in which rock particles move (vibrate) ⟵⟹

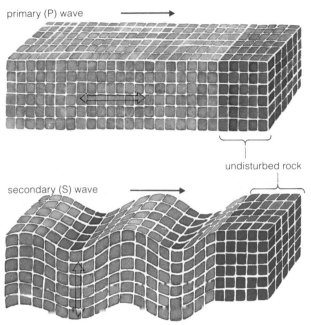

primary (P) wave

secondary (S) wave

undisturbed rock

Love wave

Rayleigh wave

undisturbed rock

Seismic waves that travel through the Earth   **Seismic waves that travel across the surface of the Earth**

The speed of seismic waves depends on the density of the rock they are travelling through. If the density rises, the waves travel faster. The chart on the right shows how the speeds of P-waves and S-waves change with depth, down to 300 km.

1 Look at the seismogram on the opposite page. Which travel faster, P-waves or S-waves? Explain how you can tell.
2 From the amplitude of the seismogram waveform, decide which waves carry the most seismic energy.
3 Study the diagrams (above left) showing how P-waves and S-waves move.
a Which is a longitudinal wave? How can you tell?
b Which is a transverse wave? How can you tell?
4 Study the diagrams (above right) showing how two different types of surface wave move.
a Which type is a transverse wave?
b How does the other surface wave travel?
5 Look at the chart on the right.
a Describe what happens to the speed of P-waves and of S-waves between 0 and 50 km depth.
b What does your last answer tell you about the density of the rock between 0 and 50 km depth?
c At what depth does there appear to be a boundary between rocks of different densities?

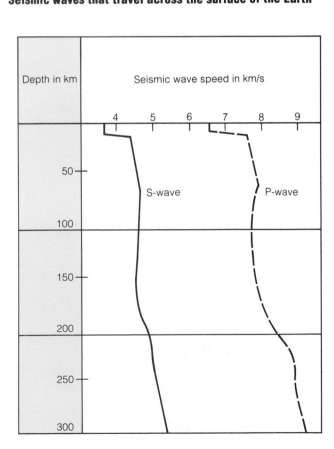

145

3.20 Looking into the Earth

The Earth is 12 800 km across, yet the deepest borehole reaches only about 15 kilometres down. So what is the inside of our planet like?

Some clues come from density measurements. Knowing the gravitational field strength and radius, scientists have calculated that the Earth's average density is 5500 kg/m³. As the density of the surface rocks is only about 2800 kg/m³, there must be much denser material towards the centre to give the higher average figure.

Evidence from seismic waves

Some clues about the Earth's structure come from records of P-waves and S-waves, as illustrated in the diagram below:

- From their travel times, seismic waves cannot be travelling through the Earth in straight lines.
- No S-waves arrive at areas on the opposite side of the Earth from the source of the seismic waves. Something blocks them.
- P-waves arrive on the far side of the Earth, but not in a ring-shaped shadow zone.

Scientists explain these observations as follows:

- In the Earth, density increases with depth. This changes the speed of seismic waves so that they are refracted (bent) and follow curved paths.
- The Earth has a central **core**. The outer part must be molten because S-waves are blocked, and S-waves cannot travel through liquids.
- The pattern of P-wave refraction suggests that the core has a density of 11 000 kg/m³ or more.

In 1914, the size of the core was measured by Bene Gutenburg using seismic wave tracing. Its radius is 3470 km. In 1936, Inge Lehmann studied how seismic waves travel through the core. She found that, 2250 km down into the core, it becomes solid, with a density of over 13 000 kg/m³. The core is probably mainly iron.

In 1909, Andrija Mohorovicic found that seismic waves are reflected at a boundary a few dozen kilometres below the surface. This boundary is called the **Mohorovicic discontinuity** or **Moho**. Above it are the rocks making the **crust**, and below it is the **mantle**.

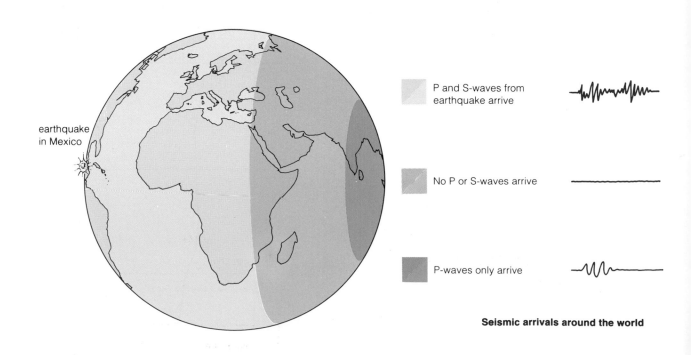

earthquake in Mexico

P and S-waves from earthquake arrive

No P or S-waves arrive

P-waves only arrive

Seismic arrivals around the world

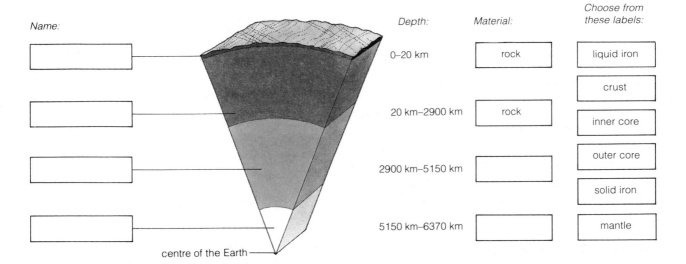

Name:

Depth:

Material:

Choose from these labels:

Depth	Material
0–20 km	rock
20 km–2900 km	rock
2900 km–5150 km	
5150 km–6370 km	

liquid iron

crust

inner core

outer core

solid iron

mantle

centre of the Earth

The structure of the Earth

1 Copy the diagram of the Earth's interior above, adding the labels to the blank boxes.

2 Copy the diagram on the right.

a Label your diagram to show the following: *crust   mantle   core*

b Seismic waves from the earthquake focus were recorded at two stations, X and Y. Which record was made at station X, and which at station Y? Explain.

3 The graph below shows the speed of seismic waves at depths in the Earth down to the centre.

a Which graph represents P-wave speeds? Why?

b Line A drops to zero speed at about 3000 km. Explain this behaviour.

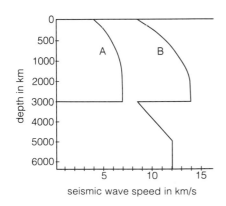

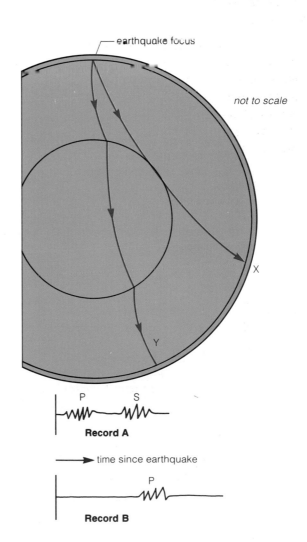

earthquake focus

not to scale

Record A

time since earthquake

Record B

Questions on Section 3

1

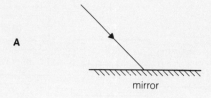

A

mirror

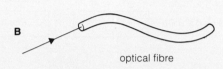

B

optical fibre

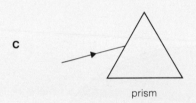

C

prism

a Copy and complete the diagrams above to show what will happen to the rays.
b Which of the above is an example of:
  i refraction;
  ii total internal reflection.

2 This diagram shows the electromagnetic spectrum, but two types of radiation are missing from it.

radio waves	micro- waves	A	light	ultra- violet	B	gamma rays

a Which radiation should be in region A?
b Which radiation should be in region B?
c Which type of electromagnetic radiation:
  i has the shortest wavelength?
  ii has the lowest frequency?
  iii will diffract round hills?
  iv gives people a sun-tan?
d Which colour of visible light has:
  i the longest wavelength?
  ii the shortest wavelength?

3 In sunny countries, some houses have a solar heater on the roof. It warms up water for the house. The diagram below shows a typical arrangement.

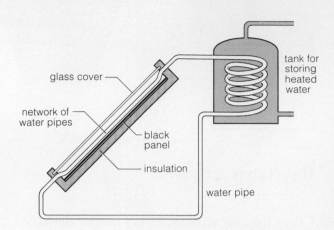

glass cover

network of water pipes

black panel

insulation

tank for storing heated water

water pipe

a Why is the panel in the solar heater black?
b Why is there an insulating layer behind the panel?
c How does the water in the tank get heated?
d Why does the solar panel work best if placed at an angle, rather than flat or upright?
e On average, each square metre of the solar panel above receives 1000 joules of energy from the Sun every second. Use this figure to calculate the power input (in kW) of the panel if its surface area is 2 m².
f The solar heater in the diagram has an efficiency of 60% (it wastes 40% of the solar energy it receives). What area of panel would be needed to deliver heat at the same rate, on average, as a 3 kW electric immersion heater?
g i What are the advantages of using a solar heater instead of an immersion heater?
  ii What are the disadvantages?

4

A Red book	B Green book	C White book

Answer each of the following by choosing A, B, or C above.
a Which of the above will absorb green and blue light, but not red?
b Which of the above will reflect red, green, and blue light?
c Which of the above will look black when viewed in green light?

5 a The diagram below shows rays of light entering someone's eye. The person is long-sighted, and the rays are not being focused correctly.

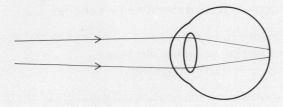

  i Where should the rays meet if the person is to see clearly?

  ii What type of spectacle lens does the person need to solve the problem?

  iii Redraw the diagram to show how, with this spectacle lens in front, the eye can focus the rays correctly.

b Describe three ways in which the human eye is similar to a camera.

6 A microphone is connected to an oscilloscope (CRO). When three different sounds, A, B, and C, are made in front of the microphone, these are the waveforms seen on the screen:

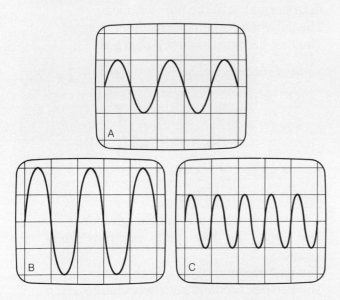

a Comparing sounds A and B, how would they sound different?

b Comparing sounds A and C, how would they sound different?

c Which sound has the highest amplitude?

d Which sound has the highest frequency?

e Sound A has a frequency of 220 Hz. If the speed of sound is 330 m/s, what is the wavelength of sound A?

f What is the frequency of sound C?

7 a Sound X: frequency 10 000 Hz.
  Sound Y: frequency 30 000 Hz.
  Upper limit of human hearing: 20 000 Hz.

  i What is the upper limit of human hearing in kHz?

  ii Which of the above sounds is an example of ultrasound?

b Ultrasound can travel through some human tissues and can be reflected by different layers inside the body.

  i Describe one example of how ultrasound is used in hospitals.

  ii For producing medical images, why do doctors prefer to use ultrasound if they can, rather than X-rays?

  iii Describe one example of the industrial use of ultrasound.

c i What is meant by *resonance*?

  ii Give one example of where resonance is a nuisance.

  iii Give one example of where resonance is useful.

8 When there is an earthquake, two types of seismic waves travel through the Earth.
P-waves are longitudinal waves.
S-waves are transverse waves.

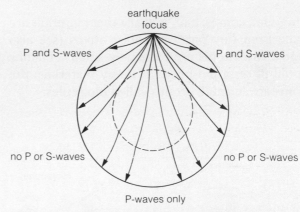

a What is the difference between *longitudinal* waves and *transverse* waves? Give one example of each type, other than seismic waves.

b Why do seismic waves travel through the Earth in curved paths?

c Why are no S-waves detected on the opposite side of the Earth from the earthquake focus?

d Explain how, by comparing signals from several monitoring stations, it is possible to work out where an earthquake has taken place.

4.1 Electric charge

It makes cling-film stick to your hands, and dust stick to a TV screen. It causes crackles and sparks when you comb your hair. It can even make your hair stand on end.

Where charge comes from

Cling-film, combs, hair, and all other materials are made from tiny particles called **atoms** (see also page 198). Atoms are extremely small – billions would fit on a pin-head. In many materials, the atoms are in small groups called **molecules**.

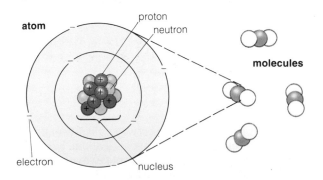

Atoms all have electric charges inside them.
In the centre of each atom there is a **nucleus**. This is made up of **protons** and **neutrons**. Even tinier particles orbit round this nucleus. These are **electrons**.

Protons and electrons both carry an electric charge. But the charges are of opposite types:

Electrons have a negative (−) charge.

Protons have a positive (+) charge, equal in size to the charge on the electron.
Neutrons have no charge.

Normally, atoms have equal numbers of electrons and protons. So the − and + charges cancel each other out. But electrons don't always stay attached to atoms. They can be removed by rubbing.

Charging by rubbing

If two materials are rubbed together, electrons may be transferred from one to the other. This upsets the balance between + and −.

A polythene comb is pulled through hair.
The polythene pulls electrons from atoms in the hair.
This leaves the polythene with more electrons than normal, and the hair with less.
The polythene becomes *negatively* charged.
The hair becomes *positively* charged.

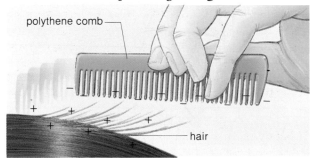

A Perspex comb is pulled through hair.
This time, the hair pulls electrons from the Perspex.
The hair becomes *negatively* charged.
The Perspex becomes *positively* charged.

Forces between charges

Like charges repel
Hold two strips of cling-film together at one end. Charge them up by pulling them between your fingers.

Both strips have the same type of charge on them. They try to push each other apart.

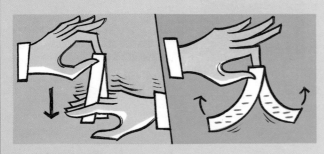

Unlike charges attract
Pull a piece of cling-film across your hand. Both become charged up. But the charges are opposite. The cling-film is attracted to your hand.

Charge attraction

A positively-charged comb is put just above a small piece of kitchen foil. Electrons in the foil are pulled upwards. This makes the top end of the foil negative. But it leaves the bottom end short of electrons, and therefore positive.

The comb attracts the negative end of the foil strongly, because it is close. It repels the positive end ... but less strongly because it is further away. The attraction wins. The foil is pulled to the comb.

This is an example of something charged (a comb) attracting something uncharged (foil). The

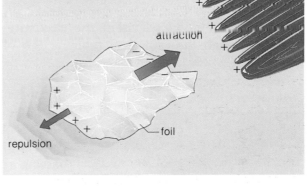

charges which appear on the foil are called **induced** charges.

1 Say whether the things below will attract each other, repel each other, or do neither:

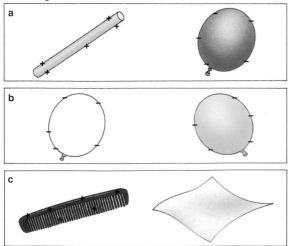

2 A balloon becomes negatively charged when rubbed against someone's sleeve.

Explain:
a how it becomes charged;
b why it will then stick to a wall.

4.2 Charge on the move

conductor

insulator

When electric charge passes through these, they can give you anything from heat to moving pictures:

When you switch on a television, the 'electricity' passing through the cable is actually a flow of electrons. The flow is called a **current**.

Electrons flow easily through the copper wire in the cable. Copper is a good **conductor**. But the electrons can't pass through the PVC plastic coating round the wire. PVC is an **insulator**.

Conductors

	In conductors, some electrons aren't very tightly held to their atoms. They can move through the material by passing from atom to atom.
	
	Good conductors of electricity are also good conductors of heat (see page 90).
		Examples
		
		metals:
		especially
		   silver
		   copper
		   aluminium
		
		carbon

Semiconductors 	Semiconductors behave like insulators when cold. When warm, they become poor conductors.	*Examples* silicon germanium

Insulators

	In insulators, the electrons are all tightly held to atoms.
	
	Insulators can be charged by rubbing. If electrons are gained or lost, they can't flow back through the material.
		Examples
		
		plastics
		e.g. PVC
		   polythene
		   Perspex
		
		glass
		rubber

Charge movers

Cells and batteries are a useful source of electric charge. They change chemical energy into electrical energy.

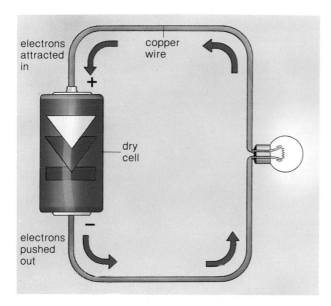

This is a **dry cell**, connected to a light bulb by two copper wires. In the cell, chemical reactions move electrons to the outside case. This makes the case negative (−) and leaves the central terminal positive (+). Electrons are pushed out of the − and attracted round to the +. As they pass through the filament of the bulb, they make it so hot that it gives out light.

When the chemicals in the cell have been used up, no more electrons can be pushed out. The cell is then 'flat'.

1   Battery Charge Current Insulator
    Conductor Cells Semiconductor

Which word describes each of the following?
a A material through which electrons can flow.
b A flow of electrons.
c A material which doesn't allow a current to pass through.
d A collection of cells.
e A material which acts as an insulator when cold, but conducts when warm.

2 Mike uses a metal comb, Rachel uses a nylon one. Explain why one of them may see sparks when they comb their hair, but the other one won't.

A **battery** is made by joining several cells together. A battery really means a group of cells, though the word is often used for just one cell.

Moving charge through liquids

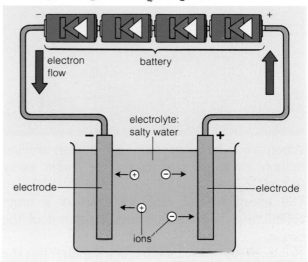

The salty water above is conducting a current. This can happen in any liquid which contains charged particles called **ions**. Ions are atoms, or groups of atoms, which have gained or lost electrons.

Positive (+) ions are pulled towards the negative (−) electrode (plate), where they collect electrons. Negative (−) ions are pulled to the positive (+) electrode, where they deliver electrons. Overall, the result is a transfer of electrons from one electrode to the other. The process, which causes chemical changes, is called **electrolysis**. New ions are formed as it takes place.

3 The chart gives information about a cassette player and the cells needed to make it work.

Number of cells needed	6
Cost of each cell	50p
Energy stored in each cell	10 000 joules
Energy used by cassette played in 1 hour	20 000 joules

a What is the total energy stored by the cells?
b For how long will the cassette player run on one set of cells?
c What is the cost of running the cassette player for one hour?

4.3 A simple circuit

When you dry your hair, 20 million million million electrons pass along the cable every second.

The hairdryer and cable are part of a huge conducting loop which passes right out of the house.

The loops on this page are much smaller. But the principles are just the same.

Current

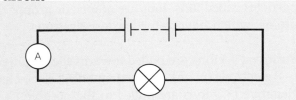

A bulb, wire, meter and battery – all drawn using electrical symbols.

The complete loop is known as a **circuit**.
The meter is measuring the flow of charge.
The meter is called an **ammeter**.
The flow of charge is called a **current**.

The unit of current is the ampere (A).

A current of 1 ampere means that about 6 million million million electrons are flowing round the circuit every second.

Typical current sizes	
Current through ...	
.. a small torch bulb	0.2A
.. a hairdryer	3A
.. a car headlight bulb	4A
.. an electric kettle	10A

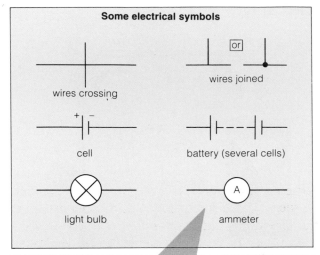

Some electrical symbols

wires crossing

wires joined

cell

battery (several cells)

light bulb

ammeter

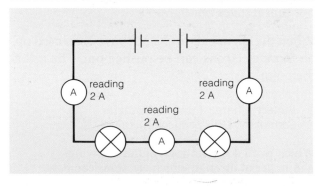

reading 2 A   reading 2 A   reading 2 A

This circuit has three ammeters and two bulbs in it. When electrons leave the battery, they flow through each of the ammeters in turn. So the readings are all the same. In a simple circuit, **the current through every part is the same.**

Putting ammeters in the circuit doesn't affect the current. As far as the circuit is concerned, the ammeters are just like pieces of connecting wire.

Which way?

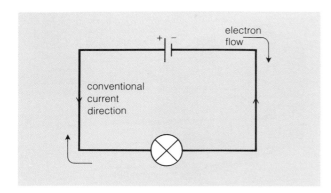

Some circuit diagrams have arrowheads marked on them. These don't show a flow. They just give the direction from positive ($+$) to negative ($-$) round the circuit. This is called the **conventional current direction**. The electrons actually flow the opposite way.

Current and charge

If a current is flowing, then electric charge is passing round a circuit.
Amounts of charge are measured in **coulombs**:

If a current of	flows for	then the charge passing is
1 ampere	1 second	1 coulomb
2 amperes	1 second	2 coulombs
2 amperes	3 seconds	6 coulombs
... and so on.		

You can use an equation to calculate charge:

charge = **current** × **time**
(coulombs) (amperes) (seconds)

Use it to check the examples above.

If you think of a current as a flow of charge, then

This current	*Means* this flow of charge
1 ampere	1 coulomb every second
2 amperes	2 coulombs every second
... and so on	

1 What is the reading on each of these ammeters?

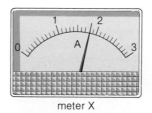
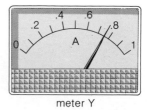

meter X   meter Y

2 Copy the diagram below.

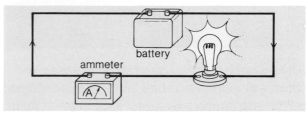

a What do the arrows on this diagram show?
b Mark in the positive and negative terminals of the battery.
Mark in the direction of electron flow, using an arrow alongside the wire.
c Redraw the diagram using the correct electrical symbols.

3 The current through the bulb is 3 A.

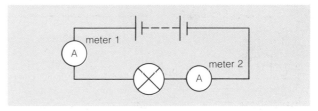

What is the current through the battery?
What is the current through meter 1?
What is the current through meter 2?

4

Appliance	Time switched on in s	Current in A
Electric drill	20	2
Food mixer		1
Hairdryer	8	

The electrical appliances in the chart were all switched on for different times.
a How much charge was taken by the electric drill?
b If the food mixer took the same charge as the electric drill, how long was it switched on for?
c If the hairdryer took the same charge as the other two, what current was flowing through it?

4.4 Voltage

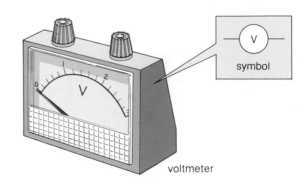

symbol

voltmeter

Anyone who attacks this fish is likely to get a shock – in more ways than one. When an electric eel senses danger, it turns itself into a living battery – pushing out electrons with nearly double the energy of those from a mains socket.

Energy from a battery

When electrons are pushed out of a battery, they carry energy with them.

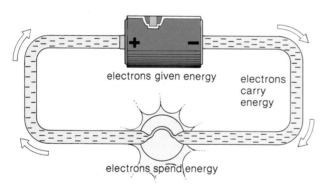

electrons given energy   electrons carry energy

electrons spend energy

In the circuit, the electrons spend all their energy passing through the bulb. The energy is changed into heat and light. When the electrons reach the battery again, all their energy has gone.

Battery voltage

Some batteries give electrons more energy than others. The higher the **voltage**, the more energy is given to each electron.

Voltage is also known as **potential difference (PD)**. The unit of PD is the **volt (V)**.
Voltage is measured by connecting an instrument called a **voltmeter** across the battery terminals. The voltage produced inside a battery is called the **electromotive force (EMF)** of the battery.

Voltages round a circuit

Three bulbs connected to a 12 volt battery. The battery gives the electrons energy.

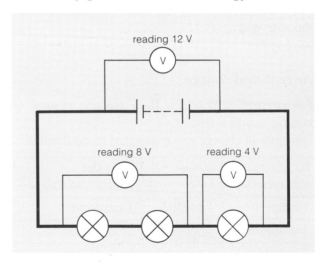

reading 12 V

reading 8 V   reading 4 V

The electrons spend some of their energy in the first bulb, some in the second, and the rest in the third.
Connect a voltmeter across any of the bulbs and it shows a reading. The higher the voltage, the more energy each electron spends as it passes through that part of the circuit.

Between them, the bulbs give out all the energy supplied by the battery:
The voltages across the bulbs add up to equal the battery voltage.

Connecting a voltmeter has almost no effect on the current flowing in the circuit. As far as the circuit is concerned, the voltmeters might as well not be there.

Cells in series and parallel

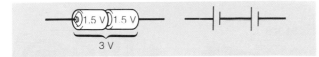

3 V

These cells are connected in **series**.
They give twice the voltage of a single cell.

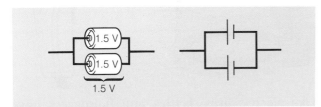

1.5 V

These cells are connected in **parallel**.
They give the same voltage as a single cell.
But they last twice as long as a single cell.

Volts, coulombs, and joules

There is an exact link between voltage, charge and energy.

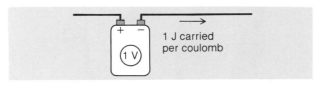

1 J carried per coulomb

1 V

Voltage across cell: 1 volt.
This cell gives 1 joule of energy to every coulomb of charge it pushes out.

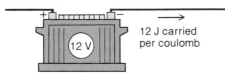

12 J carried per coulomb

12 V

Voltage across battery: 12 volts.
This battery gives 12 joules of energy to every coulomb of charge it pushes out.

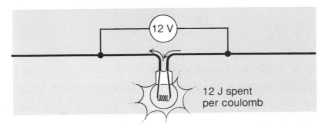

12 V

12 J spent per coulomb

Voltage across bulb: 12 volts
12 joules of energy are spent by every coulomb of charge passing through.

1 In which section of this circuit do the electrons have:
a most energy; **b** least energy?

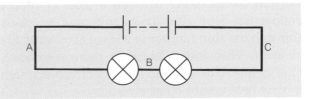

What happens to the energy they lose?
2 What is the voltage across each arrangement of cells?

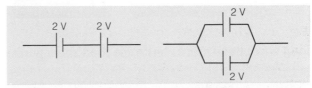

3 a What is the reading on the voltmeter across bulb B?

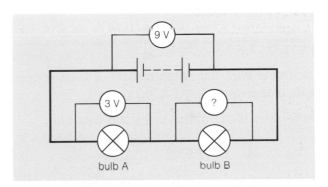

bulb A   bulb B

b How much energy does the battery give each coulomb of charge it pushes out?
c How much energy is spent by each coulomb of charge as it passes through bulb A?
4 To answer this question, you may need to look up information in the previous section.

	A Battery of dry cells	B car battery	C watch battery
EMF in V	15	12	1.5
Maximum current in A	6	100	0.01

a Which battery can push out the most electrons every second?
b Which battery pushes out electrons with most energy?
c How much charge can the car battery push out in 10 seconds?
d How much energy can the car battery deliver in 10 seconds?

4.5 Resistance

Current passes easily through a piece of copper connecting wire. But it doesn't pass so easily through the thin nichrome wire of an electric fire element. This wire has much more **resistance**. Energy has to be spent to force electrons through it. And heat comes off as a result.

All conductors have some resistance. But:

long wires have more resistance than short wires;

thin wires have more resistance than thick wires;

nichrome wire has more resistance than copper wire of the same size.

Resistance is calculated using this equation:

$$\text{resistance} = \frac{\text{voltage}}{\text{current}}$$

The unit of resistance is the **ohm** (Ω).

For example:

If there is a voltage of 12 volts across this piece of nichrome, then a current of 4 amperes flows through. So:	If there is a voltage of 12 volts across this piece of nichrome, then a current of 2 amperes flows through. So:
$\text{resistance} = \dfrac{12}{4} \text{ ohms}$ $= 3 \text{ ohms}$	$\text{resistance} = \dfrac{12}{2} \text{ ohms}$ $= 6 \text{ ohms}$

The *higher* the resistance, the *less* current flows for each volt across the wire.

Heaters...

Like electric fires, kettles and hairdryers have heating elements made from coils of thin nichrome wire. The wire gives off heat when a current passes through.

... and resistors

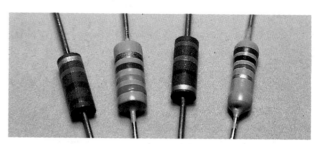

These are **resistors**. They also give off heat when a current passes through. But that isn't their job. In some circuits, they are used to reduce the current. In radio or TV circuits, they keep currents and voltages at the levels needed to make other parts work properly.

In a **variable resistor** there is a sliding contact which moves along a coil of nichrome wire. By moving the contact, you can change the resistance.

Variable resistors like this are used as volume controls in TVs and radios, and also in computer joysticks.

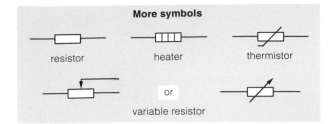

More symbols

resistor — heater — thermistor

variable resistor — or

Measuring resistance – Ohm's law

This is an experiment to measure the resistance of a length of nichrome wire when different currents are flowing through it:

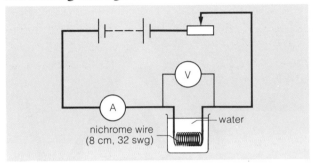

nichrome wire
(8 cm, 32 swg)
water

The voltmeter measures the voltage across the wire. The ammeter measures the current flowing through. The water keeps the wire at a steady temperature. To change the voltage across the wire, you move the sliding contact on the variable resistor. This gives the wire a different share of the battery voltage. You increase the voltage in stages, and measure the current each time. For example:

Voltage across wire in V	Current through wire in A	Voltage ÷ Current in Ω
3.0	1.0	3
6.0	2.0	3
9.0	3.0	3
12.0	4.0	3
		= resistance

Here, the voltage and current are in *proportion*. The resistance has the same value every time. Like all other metals the nichrome obeys **Ohm's law:**

The resistance of a metal conductor is the same, whatever current is flowing – provided the temperature doesn't change.

Resistance and temperature

If the temperature changes, so does the resistance.

If a *metal* is warmed, its resistance goes *up*.

If a *semiconductor* is warmed, its resistance goes *down*. Carbon behaves in the same way.

This is a **thermistor**, made from semiconducting materials. When it is warmed, its resistance falls sharply. Thermistors are used in electronic circuits which have to be 'switched' on or off by a temperature change (see page 220).

1 When a kettle is plugged into the 230 V mains, a current of 10 A flows through its element. What is the resistance of the element?

2 A piece of nichrome wire is kept at a steady temperature. Different voltages are applied across the wire, and the current measured each time. Copy the table, and fill in the missing values.

Voltage in V	Current in A	Resistance in Ω
8	2	?
4	?	?
2	?	?

3 A headlamp bulb has a filament made of tungsten metal. This is how the current through the bulb rises when the voltage across it is increased:

Voltage in V	2	4	6	8	10	12
Current in A	1.8	2.8	3.5	4.1	4.6	5.0

Plot a graph of *current* (side axis) against *voltage* (bottom axis). Use your graph to find:
a the current flowing when the voltage is 9 V;
b the resistance of the bulb when the current is 2 A;
c the resistance of the bulb when the current is 4 A;
d the highest resistance of the bulb.
Mark on your graph the point where the temperature is highest.

159

4.6 Danger! Electricity

A 132 000 volt overhead cable can push more than enough current through someone to kill them. To prevent accidents, the cables are suspended way above roof-top height. And the pylons are built so that people can't climb them. However, accidents have occurred when kite lines have touched cables.

A deadly playground

Every year, over 50 children are killed or seriously injured while playing on railway lines. With more and more track being electrified, the problem is getting worse. Contact with the live rail doesn't always kill. But it can cause serious burns as current flows through arms or legs to the ground.

Lightning doesn't always kill. But it too can cause serious burns. You are most at risk on open ground, or near an isolated tree or buildings. But the chances of being struck are still very small – much less than a big win on the football pools or lottery, for example.

It's only a 12 volt battery. So most people don't expect it to be dangerous. But if a spanner is accidently connected across the wires from the battery, the surge of current could be enough to burn you or start a fire. Wise mechanics disconnect the battery before starting work.

Fire hazards

In the home people are more at risk from electrical fires than they are from electric shocks. Here are some of the causes:

Old, frayed wiring.
Broken strands of wire can mean that a cable has a high resistance at one point. So heat is given off when current flows through. It may be enough to melt the insulation and cause a fire.

Dirty plug pins.
These give a high resistance where they connect with the socket. When a current flows through, the plug may overheat.

Too many appliances connected to one socket. If all the appliances are switched on at once, the supply cable may become overloaded.

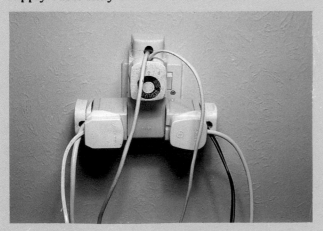

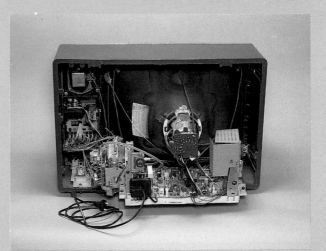

The TV is probably the most dangerous piece of equipment in the house. When a TV set is working, the voltages inside can reach 25 000 V or more. And parts inside are still live even when the set is switched off.

For safety's sake.

NEVER take the back off.

ALWAYS switch off at the plug overnight.
Unless you do this, there is still a live connection through to the set. If a fault develops, a current could flow, and a fire could start.

Can you explain why, for safety, you should disconnect the battery before working on a car engine?
Can you explain why you should NOT:

- fly kites near overhead cables?
- connect too many appliances to one socket?
- leave a television set plugged in overnight?

Try to find out why:

- bathroom lights have to be switched on and off by a pull-cord;
- extension leads shouldn't be coiled up tightly when in use;
- electric drills and food mixers are 'double insulated'.

4.7 Series and parallel

How do you run twenty dodgems from one fairground generator? Or two lamps and a hair-dryer from one mains socket? In much the same way as you run two light bulbs from one battery.

Connected to a battery, a single bulb glows brightly.

Here are two ways in which you could add a second bulb to the circuit:

Bulbs in series

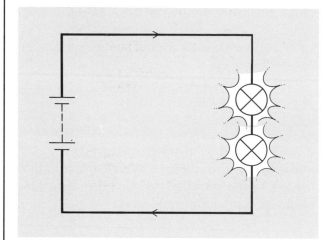

These bulbs are connected in **series**.
They have to share the battery voltage. So each glows dimly.

If one bulb is removed, the circuit is broken. The other bulb goes out.

Switches

A **switch** breaks a circuit by moving two contacts apart.
In this circuit, each bulb is controlled by a switch.
To find out which:
trace a route with your finger from one side of the battery, through a bulb, to the other side.
Your finger will pass over a switch.
This is the switch that turns the bulb on and off.

Two of the bulbs are controlled by the same switch. Can you tell which?

Bulbs in parallel

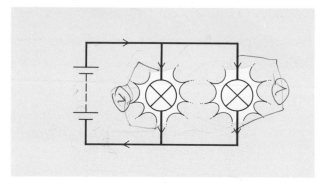

These bulbs are connected in **parallel.**
Each has direct connections to the battery.
Each gets the full battery voltage. So each glows brightly. But together, the bulbs take twice as much current as a single bulb. Energy is taken from the battery at a faster rate, so the battery goes 'flat' more quickly.

If one bulb is removed, there is still an unbroken circuit through the other bulb. So it continues to glow brightly.

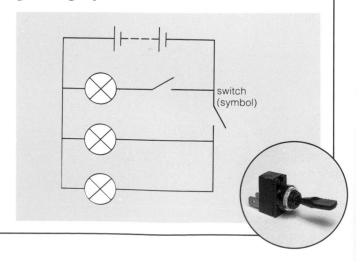

switch
(symbol)

Resistors in series

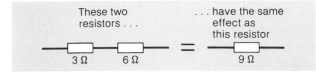

Two resistors in series.
Together, they give a higher resistance than either resistor by itself. The effect is the same as joining two short lengths of nichrome wire together to make a longer length.

To find the combined resistance, just add up the resistance values:

combined	=	first	+	second
resistance		resistance		resistance

The rule works for three or more resistances as well.

If one bulb breaks they all go off. What does this tell you about the way that these lights are connected?

Resistors in parallel

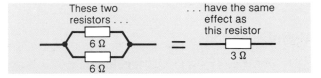

Two resistors in parallel.
Together, they give a lower resistance than either resistor by itself. The effect is the same as putting two pieces of nichrome wire side by side. They behave like a wider piece of wire.

If the two resistances are the *same*, the combined resistance is *half* a single resistance.

If the two resistances are different, you have to use an equation find the combined resistance:

$$\frac{\text{combined}}{\text{resistance}} = \frac{\text{first resistance} \times \text{second resistance}}{\text{first resistance} + \text{second resistance}}$$

For example:

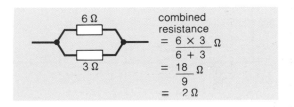

$$\text{combined resistance} = \frac{6 \times 3}{6 + 3}\ \Omega$$
$$= \frac{18}{9}\ \Omega$$
$$= 2\ \Omega$$

1 The chart give you information about three different sets of bulbs: A, B, and C. In each case, say whether the bulbs are arranged in SERIES or in PARALLEL. Then copy and complete the last column.

	Power source	Bulbs connected	Voltage across each bulb	Effect of removing one bulb
A	230 V mains	3 ceiling bulbs	230 V	others stay ON
B	230 V mains	20 Christmas tree bulbs	11.5 V	?
C	12 V battery	2 headlamp bulbs	12 V	?

2 Match the resistors!

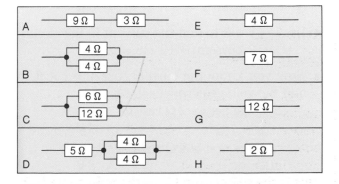

Somewhere in the right-hand column is a resistor with the same resistance as a combination in the left-hand column. Find the matching pairs.

4.8 Solving circuits

Useful equations

This equation: **resistance** $= \dfrac{\textbf{voltage}}{\textbf{current}}$

can be written using symbols: $R = \dfrac{V}{I}$

where **R** is the **resistance** in **ohms**
$\quad\quad$ **V** is the **voltage** $\quad$ in **volts**
and $\quad$ **I** is the **current** $\quad$ in **amperes**

You can rearrange the equation in two ways:

$$I = \frac{V}{R} \text{ and } V = I \times R$$

These are useful if you know the resistance, but need to find the current or voltage.

This triangle gives you all three equations. If you want the equation for *I*, just cover up *I*, and so on....

For example:
A current of 2 A flows through a 6 Ω resistor. To find the voltage across the resistor:

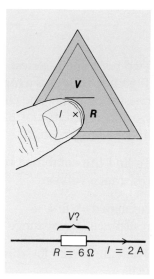

Select the equation for *V* and fill in the values of *I* and *R*:
$$V = I \times R$$
$$= 2 \times 6$$
$$= 12\,V$$

When resistors are in series ...

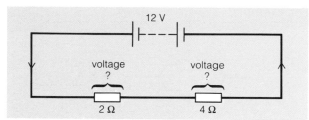

the current is the same through each.

But the voltage is shared.

Problem: to find out the voltage across each resistor in this circuit:

- Find the total resistance in the circuit:
$\quad\quad$ resistance $= 2\,\Omega + 4\,\Omega = 6\,\Omega$

- Use $I = V \div R$ to find the current in the circuit:
$\quad\quad$ current $= 12 \div 6 = 2\,A$

- Now you know the current, use $V = I \times R$ to find the voltage across each resistor:
$\quad\quad$ voltage across $2\,\Omega$ resistor $= 2 \times 2$
$\quad\quad\quad\quad\quad\quad\quad\quad\quad\quad = 4\,V$
$\quad\quad$ voltage across $4\,\Omega$ resistor $= 2 \times 4$
$\quad\quad\quad\quad\quad\quad\quad\quad\quad\quad = 8\,V$

- Check your answers:
The voltages across the resistors should add up to equal the battery voltage (12 V). Do they?

When resistors are in parallel ...

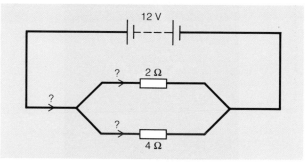

the voltage is the same across each.

But the current is shared.

Problem: to find the currents flowing in the different sections of this circuit:

- Use $I = V \div R$ to find out the current through each resistor:
$\quad\quad$ Both resistors have 12 V across them, so
$\quad\quad$ current through $2\,\Omega$ resistor $= 12 \div 2$
$\quad\quad\quad\quad\quad\quad\quad\quad\quad\quad\quad = 6\,A$
$\quad\quad$ current through $4\,\Omega$ resistor $= 12 \div 4$
$\quad\quad\quad\quad\quad\quad\quad\quad\quad\quad\quad = 3\,A$

- Add the currents together to find the current in the main circuit:
$\quad\quad$ current in main circuit $= 6\,A + 3\,A$
$\quad\quad\quad\quad\quad\quad\quad\quad\quad\quad = 9\,A$

Simpler than it looks . . .

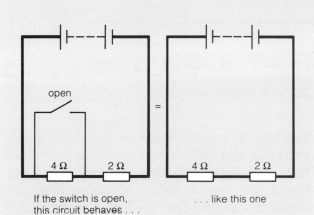

The meters don't affect the circuit. This circuit behaves . . .

. . . like this one

If the switch is open, this circuit behaves . . .

. . . like this one

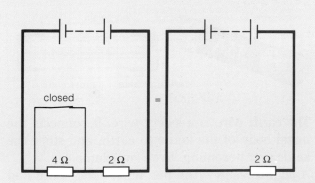

closed

But if the switch is closed, all the current takes the 'short circuit' route through the switch. It's just as if the 4 Ω resistor wasn't there.

1 In each of the following, the *resistance, voltage* or *current* needs to be calculated. Find the missing value:

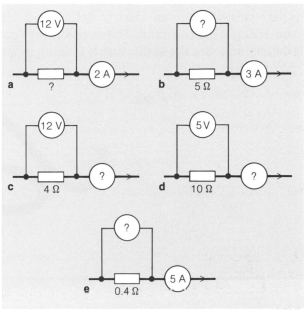

2

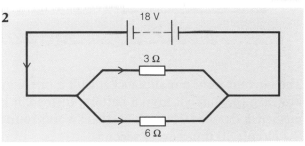

a In the circuit above, what is the current through the 6 Ω resistor?
b What is the current through the 3 Ω resistor?
c What current flows from the battery?
d Redraw the circuit, replacing the two parallel resistors by a single resistor.
If the current from the battery is the same as before, what is the resistance of this resistor?
e Use the equation for parallel resistances on page 163 to calculate the combined resistance of the resistor in the diagram above.

3

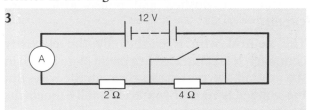

In the circuit above, what is the reading on the ammeter when the switch is **a** open **b** closed?

4.9 Mains electricity

When you plug in an electric kettle, you are connecting it into a circuit. The circuit hasn't got a battery in it. But the mains supply is doing much the same job.

AC supply (symbol)

supply cable

Live

230 V AC

Neutral

Earth (symbol)

fuse

insulated cable

The current from a mains socket isn't a one-way flow like the current from a battery. Instead, it is pushed and pulled forwards and backwards round the circuit 50 times every second.

The current is known as **alternating current** or **AC.**

The **mains frequency** is 50 Hz.

Power-stations supply AC because it is easier to generate than one-way **direct current (DC)**.
In Britain, the supply voltage is 230 V.

The connecting wires to the kettle are insulated. They are all contained in a single cable or 'flex'.

The live wire goes alternately − and + as electrons are pushed and pulled around the circuit.

The neutral wire is earthed by the elecricity company. It is connected to a metal plate buried in the ground. Current passes through the wire. But the voltage is zero. If you accidentally touch the neutral wire, you should not get a shock.

The switch on the mains socket is fitted in the live wire. This is to make sure that none of the wire in the flex is live when the switch is turned off.

The fuse is a short piece of thin wire which overheats and melts if too much current flows through it. If a fault develops, the fuse 'blows' and breaks the circuit before anything else can overheat and catch fire. The fuse is inside a small cartridge in the plug. Like the switch, it is placed in the live wire.

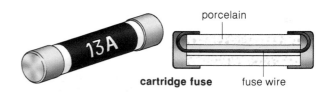

porcelain

cartridge fuse fuse wire

The earth wire is a safety wire. It connects the metal body of the kettle to earth, and stops the kettle ever becoming 'live'. For example:
A fault develops – the live wire works loose and touches the metal body of the kettle.
Results – a large current flows to earth, and blows the fuse. So the kettle isn't dangerous to touch.

Your hairdryer or radio probably doesn't have an earth wire connected to it. This is because it has an insulating plastic case, and not a metal one.

Three-pin plugs

Plugs are a simple and safe way of connecting things to a mains circuit. In Britain, the square-pin fused plug is the most commonly used type:

When wiring a plug, check that:

- The three wires are connected to the correct terminals:

  brown to **Live**
  blue to **Neutral**
  yellow and green to **Earth**

- There are no loose strands of wire.
- The cable is held firmly by the grip.
- A fuse of the correct value is fitted.

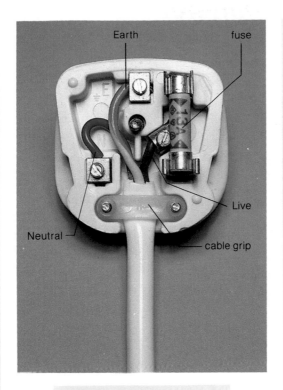

Earth · fuse · Live · cable grip · Neutral

If a fuse blows

- Switch off at the socket and pull out the plug.
- Don't fit a new fuse until the fault has been put right.

Choosing a fuse

Plugs are normally fitted with 3 A or 13 A fuses. The value tells you the current needed to 'blow' the fuse.

If a TV takes a current of 0.5 A, its plug should be fitted with a 3 A fuse.

If a kettle takes a current of 10 A, its plug should be fitted with a 13 A fuse.

The fuse value should always be more than the actual current, but as close to it as possible. The TV will still work with a 13 A fuse fitted. But it might not be safe. If something goes wrong, the circuits could overheat and catch fire, without the fuse blowing.

---

1 LIVE NEUTRAL EARTH
Which of these wires:
a has a brown covering;
b is a safety wire;
c goes alternately + and −;
d has a blue covering;
e has a yellow and green covering;
f forms part of the circuit, but has no voltage on it.
2 Copy and complete the table to show whether a 3 A or 13 A fuse should be fitted to the plug connected to each applicance. The first is done for you.

Applicance	Current taken in A	Fuse value in A
Radio	0.1	3
Hairdryer	4	
Refrigerator	0.5	
Cassette player	0.2	
Fan heater	12	
Food mixer	2	

3 This circuit has been wrongly wired.

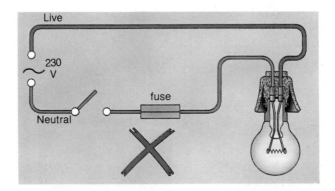

Live · 230 V · Neutral · fuse

If the bulb is taken out of its socket, the circuit isn't safe. Explain why not. Redraw the circuit, showing the correct wiring.

If your circuit were used to supply current to a metal fan heater, an earth wire should be fitted. Why should this be done?

4.10 Electrical power

They both change electrical energy into sound energy. But hers has more **power** than his.
It changes more energy every second.

Power is measured in joules per second, or **watts (W)**.

A power of 1 watt means that 1 joule of energy is being changed every second.

Typical powers

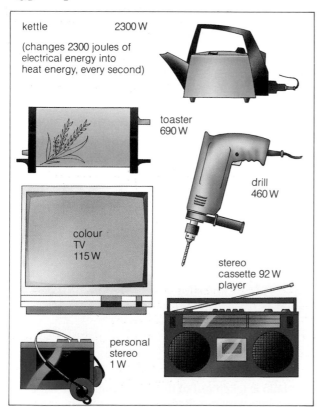

kettle   2300 W
(changes 2300 joules of electrical energy into heat energy, every second)

toaster 690 W

drill 460 W

colour TV 115 W

stereo cassette 92 W player

personal stereo 1 W

Power is sometimes given in **kilowatts**:
1 kilowatt (kW) = 1000 watts
The kettle has a power of 2.3 kW.

An equation for electrical power

You can calculate electrical power using the equation:

power = voltage × current
(watts)   (volts)   (amperes)

For example:
if a 230 V hairdryer takes a current of 2 A,
power = 230 × 2 = 460 W

● A higher voltage gives more power because each electron carried more energy.

● A higher current gives more power because there are more electrons to spend their energy every second.

Why the equation works

First, look up the meanings of *current* and *voltage* on pages 155 and 157.

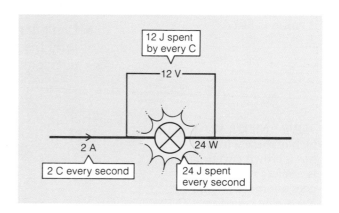

12 J spent by every C

12 V

2 A

2 C every second

24 W

24 J spent every second

This bulb has a current of 2 A flowing through it. It has a voltage of 12 V across it:
So,
2 coulombs of charge are passing through the bulb every second;
each coulomb spends 12 joules of energy as it passes through.

This means that 12 × 2 joules of energy are spent every second.
So, the power is 24 joules per second, or 24 watts.

To get this answer, you have to multiply the *voltage* by the current.

More equations

The power equation can be written using symbols:

$$P = V \times I$$

You can rearrange the equation in two ways:

$$V = \frac{P}{I} \quad \text{and} \quad I = \frac{P}{V}$$

These are useful if you know the power, but need to find the voltage or current.

More about fuses

A kettle has more power than a TV.
It takes more current from the mains.
It needs a higher value fuse in its plug:

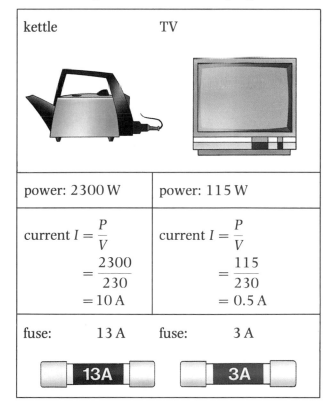

kettle	TV
power: 2300 W	power: 115 W
current $I = \dfrac{P}{V}$ $= \dfrac{2300}{230}$ $= 10\,A$	current $I = \dfrac{P}{V}$ $= \dfrac{115}{230}$ $= 0.5\,A$
fuse: 13 A	fuse: 3 A

1 Julie is setting up a lighting display in a shop window. The cable to the window can take a maximum current of 5 A. If the mains voltage is 230 V:
a what is the maximum power which can be carried by the cable?
b how many 100 W light bulbs can Julie run from the cable?

2

Mains voltage: 230 V		
A	B	C
current: 2 A	current: 3 A	current: 0.4 A

You will find these three appliances in the chart on the opposite page. Calculate the power of each one. Then work out what they are.

3

Mains voltage 230 V		
A	460	watt vacuum cleaner
B	920	watt iron
C	1150	watt fan heater
D	23	watt cassette player
E	46	watt video recorder

a What is the power of each appliance in kilowatts?
b What current is taken by each appliance?
c What fuse (3 A or 13 A) should be fitted to each plug?

4

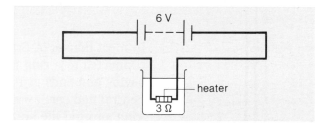

A small heater is being used to warm some water.
a What is the current through the heater?
b What is the power of the heater?

A different battery is put in the circuit. This has *twice* the voltage of the old one.
c What is the current through the heater?
d What is the power of the heater?

4.11 Circuits around the house

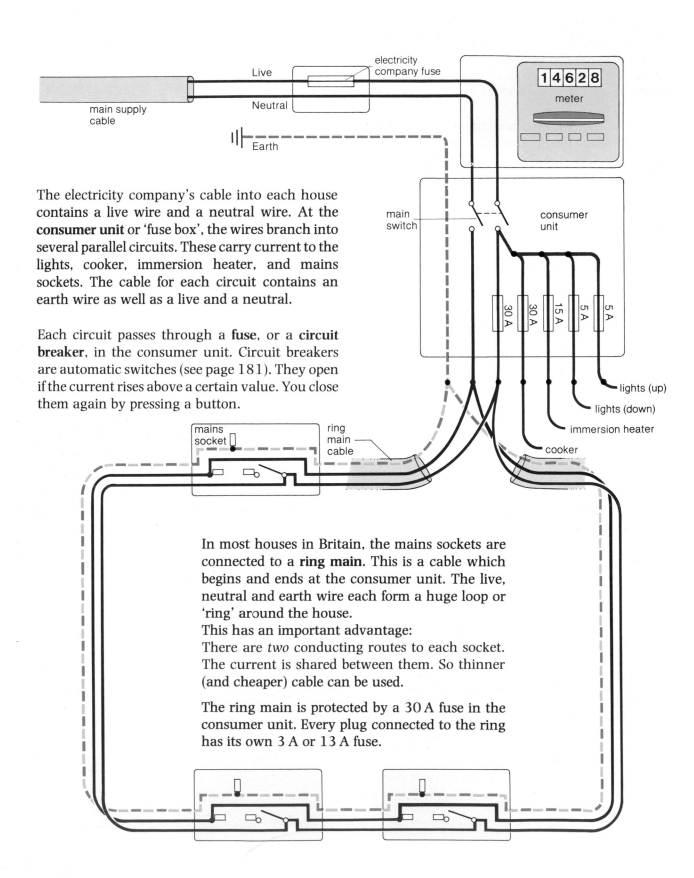

main supply cable

Live

electricity company fuse

Neutral

Earth

14628

meter

main switch

consumer unit

30 A 30 A 15 A 5 A 5 A

lights (up)

lights (down)

immersion heater

cooker

mains socket

ring main cable

The electricity company's cable into each house contains a live wire and a neutral wire. At the **consumer unit** or 'fuse box', the wires branch into several parallel circuits. These carry current to the lights, cooker, immersion heater, and mains sockets. The cable for each circuit contains an earth wire as well as a live and a neutral.

Each circuit passes through a **fuse**, or a **circuit breaker**, in the consumer unit. Circuit breakers are automatic switches (see page 181). They open if the current rises above a certain value. You close them again by pressing a button.

In most houses in Britain, the mains sockets are connected to a **ring main**. This is a cable which begins and ends at the consumer unit. The live, neutral and earth wire each form a huge loop or 'ring' around the house.
This has an important advantage:
There are *two* conducting routes to each socket. The current is shared between them. So thinner (and cheaper) cable can be used.

The ring main is protected by a 30 A fuse in the consumer unit. Every plug connected to the ring has its own 3 A or 13 A fuse.

Two-way switches

In most houses, you can turn the landing light on or off using upstairs or downstairs switches. These aren't the simple 'on-off' type. They have two contacts instead of one. They are **two-way switches**:

If the switches are both up or both down, then a current flows through the bulb. But if one switch is up and the other down, the circuit is broken. Move either switch and you reverse the effect of the other one.

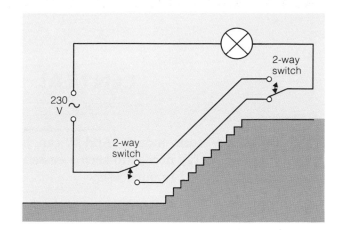

Safety first

If an accident happens ...

before you give any help:
- **switch off at the socket**
- **pull out the plug**

To prevent accidents ...

fit a **residual current circuit breaker (RCCB)**. This compares the currents in the live and neutral wires. They should be the same. If they're different, then current must be flowing to earth – perhaps through someone touching a faulty wire. The RCCB senses the differences and switches off the power before any harm can be done.

1 If an accident occurs, and someone is electrocuted, what should you do before giving any assistance?

2 When you switch on a hairdryer, the current flows through three fuses. Where are these fuses?

3 Copy and complete the diagram to show how the sockets can be connected to the consumer unit using a ring main.

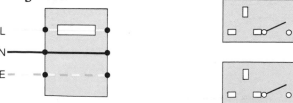

What is the main advantage of the ring main?

4 Copy and complete the diagram to show how the light bulb can be controlled by either of the switches.

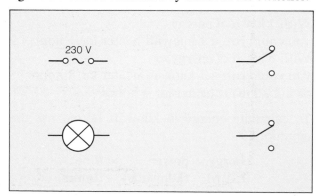

4.12 Buying electricity

UNITS USED	PRICE PER UNIT p	£
CENTRAL ELECTRICITY BOARD	VAT Registration No. 338 7449 45	
1220	10.00	122.20
QUARTERLY CHARGE		12.00
	AMOUNT INC VAT	134.20

Electrical energy costs money. And it can be expensive. The energy needed to keep a cassette recorder running for 24 hours costs:

about 2p,   about £20,
on the electricity bill   buying batteries

Working out the energy . . . in joules

A heater with a power of 1 watt (W) changes 1 joule of electrical energy into heat energy every second.
So:

With 1 joule of energy,
you could run a 1 watt heater for 1 second.

With 6 joules of energy,
you could run a 2 watt heater for 3 seconds,
or a 6 watt heater for 1 second.
And so on.

To calculate energy in joules, use the equation:

$$\text{energy} = \text{power} \times \text{time}$$
$$\text{(joules)} \quad \text{(watts)} \quad \text{(seconds)}$$

Working out the energy . . . in kilowatt hours

The 'units' on an electricity bill are units of energy called **kilowatt hours (kW h)**. The electricity company charges you a set amount for each kW h bought.

With 1 kW h of energy,
you could run a 1 kilowatt heater for 1 hour.
With 8 kW h of energy,
you could run a 1 kilowatt heater for 8 hours,
or a 2 kilowatt heater for 4 hours.

To calculate energy in kilowatt hours, use the equation:

$$\text{energy} = \text{power} \quad \times \text{time}$$
$$\text{(kW h)} \quad \text{(kilowatts)} \quad \text{(hours)}$$

The cost of drying your hair

If a 1 kW hairdryer is switched on for 15 minutes,
the power   = 1 kW
the time   = 0.25 hours
So, using the energy equation,
the energy bought = 1 × 0.25
                  = 0.25 kW h

If each kW h *or 'unit' costs* 10p,
then the total cost = 0.25 × 10
                    = 2.5p

If each kilowatt hour of energy costs 10p, (including VAT) then it will cost about . . . .
A 5p to watch TV all evening.
B 15p to bake a cake.
C 30p to wash one load of clothes.
D 240p to leave a fan heater running all day.

A bath a day

Don is horrified by his electricity bill.
Should he stop taking hot baths and buy de-odorant instead? How much is his daily bath costing him?

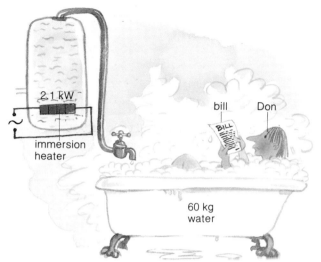

In his bath, he uses 60 kg of water.
From cold, the water has to be heated up by 30 °C.
The water has a specific heat capacity of 4200 J/(kg °C):
4200 joules of energy are needed to heat every kg through each °C.
So,
the heat energy needed = 4200 × 60 × 30
$$= 7\,560\,000 \text{ joules}$$

An immersion heater warms his water.
It has a power of 2.1 kW:
It supplies 2100 joules of heat energy every second.
But, 7 560 000 joules are needed altogether.

So, the time taken $= \dfrac{7\,560\,000}{2100}$
$$= 3600 \text{ seconds}$$
$$= 1 \text{ hour}$$

This is how long the immersion heater has to be switched on for.

The cost can be worked out from the following:
power of heater: 2.1 kW
time: 1 hour
cost of each kW h: 10p
Can you finish the calculation?

1

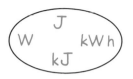

Which of these are units of energy?

2

Electric fire	
Power, in kilowatts:	1 kW
Time switched on, in hours:	1 h
Power in watts:	?
Time switched on, in seconds	?
Energy bought, in kilowatt hours	?
Energy bought, in joules	?

a Copy the table and fill in the missing values.
b If you can buy 1 kW h of energy for 10p, how many joules of energy can you buy for 10p?
3 In the chart on the opposite page, how much energy is bought in each case (A to D)? Give your answers in kW h.

4

Appliance	Power	Time
Mains radio	10 W	16 hours
Electric blanket	100 W	8 hours

Donna likes to leave her radio on all day. Her father keeps his electric blanket switched on all night. Each thinks that the other is wasting electricity. But who is adding most to the bill? Use the information in the chart to find out.
5 If each kW h of energy costs 10p, what is the cost of:
a Leaving a 2 kW fire on for 5 hours.
b Leaving a 100 W lamp on for 10 hours.
c Using a 800 W microwave oven for 15 minutes?
6 Finish the calculation on the left to find out the cost of Don's daily bath.
Don decides to take showers instead of baths.
If the water temperature is the same, but he only uses 6 kg of water each time:
a How much energy is needed to heat the water?
b How long must the 2.1 kW heater be switched on for?
c What is the cost of each shower?

4.13 Magnets

Is it worth buying? Check it with a magnet:

Ordinary steel is attracted to a magnet. But the very best quality stainless steel isn't. If the cutlery is expensive, it shouldn't do this.

If the bodywork isn't attracted to the magnet, then plastic filler has probably been used to cover up rust or crash damage.

Poles of a magnet

Iron filings cling to a bar magnet. The magnetic pull seems to come from two points near the ends. These are the **poles** of the magnet.

If you hang a bar magnet up with a piece of thread, it swings round until it lies roughly north-south. This effect gives the poles their names – the **north pole** and the **south pole**.

Pushes and pulls

Bring the ends of two identical bar magnets together and there is a force between the poles:

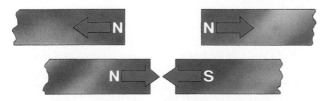

If the poles are the same, they repel (push each other apart).
If the poles are different, they attract each other.

Permanent and temporary magnets

Pieces of iron and steel *become* magnets if you place them near a magnet. The magnet **induces** magnetism in both metals. The magnet attracts the pieces of metal because the poles nearest each other are different.

When the pieces of metal are pulled away, the steel keeps its magnetism, but the iron does not. The steel has become a **permanent** magnet. The iron was only a **temporary** magnet.

Magnets inside magnets

In a piece of iron or steel, every atom is a tiny magnet. Normally, these tiny magnets point in all directions. So their effects cancel out. But when the iron or steel is magnetized, the atoms are turned so that they line up. Billions of tiny magnets then act as one big magnet.

If you hammer a magnet, atoms are thrown out of line again. Strong heating has the same effect. The magnet becomes **demagnetized**.

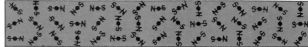

unmagnetized steel

magnetized steel

Hard or soft?

Iron and steel can both be magnetized. They are **ferromagnetic materials**. One is **hard**. The other is **soft**. But the words don't have their usual meanings.

Hard magnetic materials

These are difficult to magnetize. But they are also difficult to demagnetize. Once magnetized, they keep their magnetism.

Steel is used to make permanent magnets. **Alcomax** makes an even stronger magnet than steel.

A cassette tape is coated with tiny particles of **iron oxide**. These become magnetized when you make a recording. The strength of the magnetism varies along the tape, so that the particles form a magnetic 'copy' of the original sound waves.

Soft magnetic materials

These are easy to magnetize. But they quickly lose their magnetism.

Iron and **mumetal** are used in electromagnets because their magnetism can be 'switched' on or off (see page 168).

Most materials are **non-magnetic**. They can't be magnetized.

Many metals are non-magnetic. For example: copper, brass, aluminium, silver, gold.

---

1 Pieces of iron and steel are pulled to the ends of a magnet.

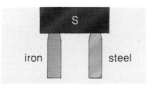

a Copy the diagram. Draw in any magnetic poles on the iron and steel.

b If the lower ends of the iron and steel start to move, which way will they move, and why?
c What happens to each of the metals when it is taken away from the magnet?

2 Which material is the odd one out in each of these lists? Why?

a	copper aluminium steel brass	**b**	steel iron iron oxide alcomax

3 Write down two ways in which a magnet could be demagnetized.

4 This is a magnetically-operated switch.

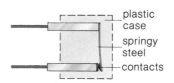

What happens when the magnet is brought close?

12 V alarm bell

12 V battery

door

This door is to be protected by a burglar alarm. Copy the diagram. Show where you would fit a magnetically-operated switch, magnet, and connecting wire so that the alarm bell will ring if the door is pushed open.

4.14 Magnetic fields

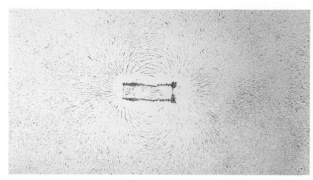

Iron filings, sprinkled on paper over a bar magnet. The filings have become tiny magnets – pulled into position by forces from the poles of the magnet.

Magnetic forces are acting in the space around the magnet. There is a **magnetic field** in this space.

The compass

Magnetic fields can be studied using a small compass. Inside the compass is a tiny magnet called a **needle**. It is on a spindle and can turn freely. The north end of the needle is a pointer.
Near a magnet, the needle is turned by forces between its poles and the poles of the magnet. The needle comes to rest with the turning forces balanced.

Plotting a field

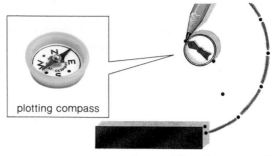

plotting compass

- Put a magnet in the middle of a piece of paper. Draw round it to mark its position.

- Place a plotting compass near one end of the magnet. Mark the position of the needle with two pencil dots.

- Move the compass so that the needle lines up with the last dot you made. Mark the position of the needle again ... and so on until you reach the magnet or the edge of the paper.

- Join up the dots. When you do this, you are drawing a **field line**.

- Repeat, starting at different points round the magnet. You can draw any number of field lines, but it's simplest not too show too many.

Field lines around a bar magnet:

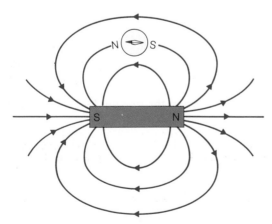

- The lines show the direction a compass needle would point. They run from the north pole of the magnet round to the south.

- The field is strongest where the lines are closest together.

Field lines between magnets:

Poles different
The field lines run from north to south.

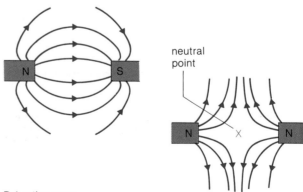

neutral point

Poles the same
At point X, the field from one magnet cancels the field from the other. X is called a neutral point. If you placed a compass at X, the needle wouldn't be turned by the magnets.

The Earth's magnetic field

The Earth has a magnetic field. It isn't very strong. No one knows what causes it. But it is rather like the field around a huge bar magnet.

If there aren't any other magnets around, a compass needle turns into line with the Earth's magnetic field.

The *north* pole of the needle points *north*. But a *north* pole is always attracted to a *south* pole. So the *south* pole of the Earth's magnet is actually in the *north*! It lies under a point in northern Canada called 'magnetic north'.

'Magnetic north' is over 1200 kilometres away from the North Pole. This is because the Earth's magnet isn't quite in line with its true north-south axis.

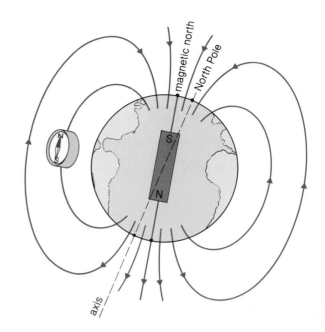

Did you know?

Seamen used to think that there was a huge magnetic mountain in northern seas. It attracted compass needles and pulled ships to their doom.

No one has ever found 'magnetic north'. It moves too fast. Every day, it travels in circles across the ice at about 5 m/s. And no one has ever found 'magnetic south' either.

Racing pigeons probably use the Earth's magnetic field to find their way home. Part of their brain acts like a compass.

---

1 Things are missing from these diagrams.

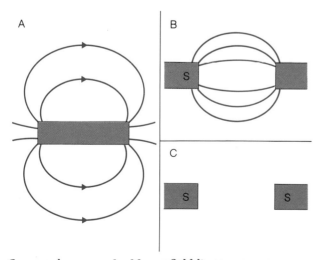

Copy each one, and add any field lines, arrows, or magnetic poles which should be there.
In which diagram could a neutral point be shown?

2

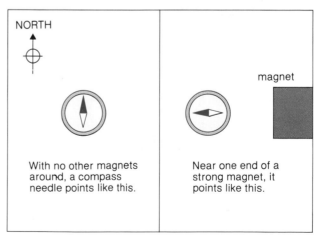

a At which end of the compass needle is there a north pole?
b What type of pole is at the end of the magnet?

4.15 Magnetism from currents

Make a recording on this tape and 45 minutes of sound becomes over 100 metres of varying magnetism along the tape. But the tape isn't magnetized by a magnet. It is magnetized by a current passing through a piece of wire.

Field around a wire

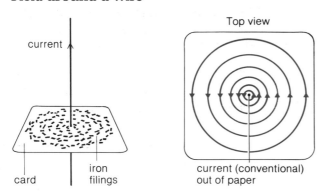

A high current flows through a wire. Iron filings are pulled into circles around it. The current produces a weak magnetic field:

- The magnetic field is strongest close to the wire.
- Increasing the current makes the magnetic field stronger.

Field around a coil

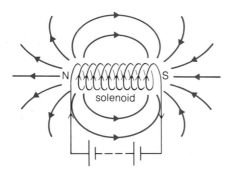

A current flows through a long coil, called a **solenoid**. A magnetic field is produced. The field is like the field around a bar magnet. The solenoid behaves as if it has magnetic poles at its ends.

- Increasing the current makes the magnetic field stronger.
- Increasing the number of turns on the coil makes the magnetic field stronger.

Rules to remember

Current flows from + to −.
This is the conventional current direction.

Magnetic field lines run from N to S.

The right-hand grip rule for field direction:

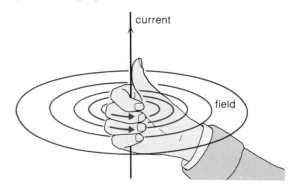

Imagine your right hand gripping the wire so that your thumb points the same way as the current. Your *fingers* curl the same way as the *field* lines.

The right-hand grip rule for poles:

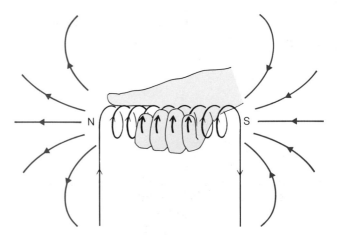

Imagine your right hand gripping the solenoid so that your fingers curl the same way as the current. Your *thumb* will then point to the *north* pole of the solenoid.

Making magnets...

A current flows through a solenoid. In the solenoid is a bar of steel. The steel becomes magnetized and makes the magnetic field much stronger than before.

When the current is switched off, the steel stays magnetized. It has now become a permanent magnet. Nearly all permanent magnets are made in this way.

...and demagnetizing them

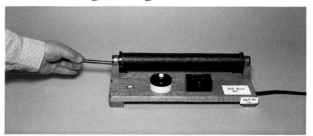

A solenoid can also be used to demagnetize a magnet. The magnet is put in the solenoid. Alternating current (AC) is passed through the solenoid. Then the magnet is slowly pulled out.

When AC is passing through the solenoid, the magnetic field keeps changing direction very rapidly. This turns the magnetic atoms in the steel out of line.

Making recordings...   ...and removing them

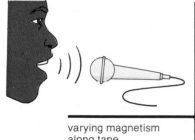

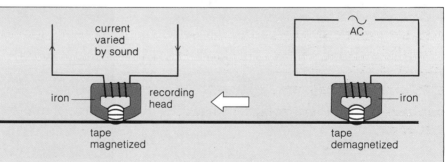

Magnetic tape passes over the recording head in a cassette recorder. The tape is magnetized by the current in the coil. As the sound varies, so does the current – and the strength of magnetism along the tape. Result: a magnetic 'copy' of the original sound waves.

New recordings can't be made until old ones have been removed. This head has AC flowing through its coil. It demagnetizes the tape passing over it – ready for the next recording.

1 This is the end view of a long length of wire. A high current is flowing through the wire. In which direction is the current flowing? – INTO the paper? – or OUT OF the paper? Copy the diagram, and show which way the other compass needles would be pointing.

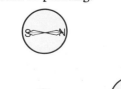

wire

2 All four coils have the same current passing through them. Which one:
a gives the weakest magnetic field;
b has a north pole at the left-hand end;
c will still give a magnetic field when the current is switched off?

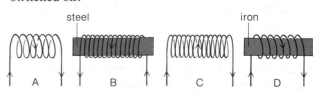

4.16 Electromagnets

Electromagnets can do all the things that ordinary magnets can do. But you can switch them on and off.

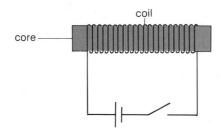

The parts of an electromagnet:

A **coil** – made from several hundred turns of insulated copper wire. The greater the number of turns, the stronger the field.

A **battery** – to supply current. The higher the current, the stronger the field.

A **core** – made from a soft magnetic material like iron. This makes the field much stronger. But its magnetism dies away as soon as the current is switched off.

Using electromagnets

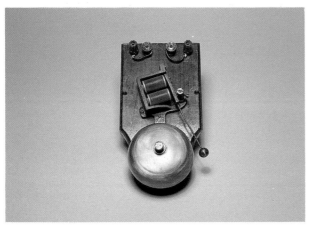

In an electric bell, an electromagnet is switched on and off very rapidly. It keeps pulling the hammer over to the gong, then releasing it.

Sorting scrap metal. Electromagnets are used to separate metals like iron and steel from other metals.

There's an electromagnet in a telephone earpiece. As the current through it varies, the pull on a metal plate varies. This makes the plate vibrate and send sound waves into your ear.

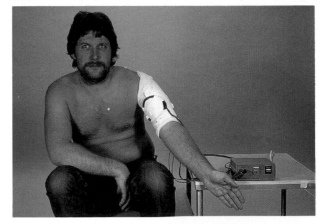

A wrap-round electromagnet. Bursts of magnetism actually help broken bones to mend more quickly. No one is quite sure why.

Switches worked by electromagnets

Problem A car starter motor takes a current of over 100 A. It has to be switched on by a lightweight switch, connected to thin cable. This can't handle the high current.

Solution Use an electromagnetic switch or **relay**. With a relay, a large current in one circuit can be switched on by a small current in another circuit.

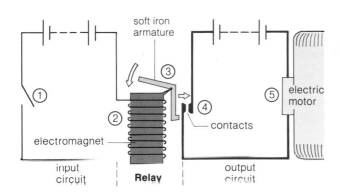

1 When the switch in the input circuit is closed:
2 The electromagnet comes ON.
3 And pulls the iron **armature** towards it.
4 This closes the contacts.
5 So the motor in the output circuit is switched ON.

The **circuit breaker** below is designed to cut off the current in a circuit if it rises above a certain value. The current flows through two contacts and also through an electromagnet. If the current gets too high, the pull of the electromagnet becomes strong enough to release the iron catch, so the contacts open.

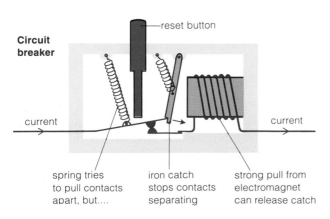

Circuit breaker

1 To answer this question, you may need information from page 178.

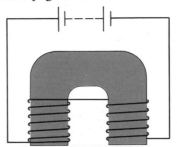

This a U-shaped electromagnet.
a Copy the diagram. Label the COIL and the CORE, and say which material each could be made from.
b Explain why steel wouldn't be a suitable material for the core.
c Explain why the wire in the coil needs to be insulated.
d Which pole is at the left-hand end of the electromagnet?
e What two changes could you make to give a stronger magnetic field?
2 In the diagram on the left, the electric motor is controlled by a relay. Below, the arrangement has been redrawn using circuit symbols.

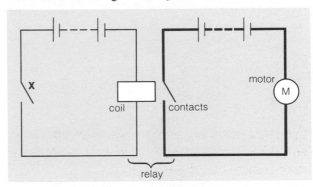

a Explain why the motor is turned on when switch X is closed.
b Phil can't see the point of the relay. It looks very complicated. He wants to know why the motor can't just be turned on and off by a simple switch in the motor circuit. What answer would you give him?
3 The diagram on the left shows a circuit breaker.
a What makes the circuit breaker switch off if the current rises above a certain value?
b What changes would you make to the circuit breaker so that it switches off at a higher current?

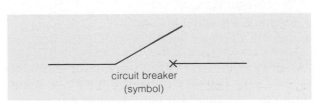

circuit breaker
(symbol)

4.17 The magnetic force on a current

Put a current into each of these, and something moves:

The loudspeaker
cone vibrates

The pointer moves
up the scale

The motor turns

The movement is caused by a force. A force is produced whenever a current flows with a magnetic field across it.

Fleming's left hand rule

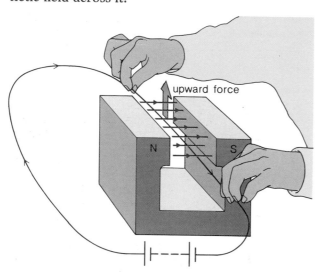

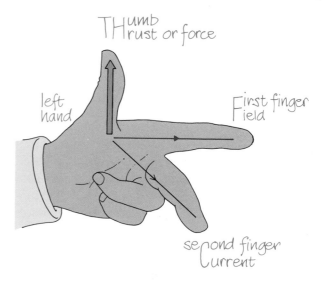

A wire is held between the poles of a magnet. When a current flows through the wire, there is an upward force on it.

The force becomes stronger if:
● the current is increased;
● a stronger magnet is used;
● there is a greater length of wire in the field.

The force isn't always upwards. It depends on the current and field directions. If the wire is in line with the field, there isn't any force at all.

This is a rule for working out the direction of the force when a current is at right angles to a magnetic field.

Hold the thumb and first two fingers of your left hand at right angles. Point your fingers as shown, and your thumb gives the direction of the force.

When you use the rule, remember:
the current direction is from + to −;
the field lines run from N to S.

Turning a coil

A coil lies between the poles of a magnet. A current flows through the coil. The current flows in opposite directions along the two sides of the coil. So one side is pushed *up*, and the other side is pushed *down*. There is a turning effect on the coil.

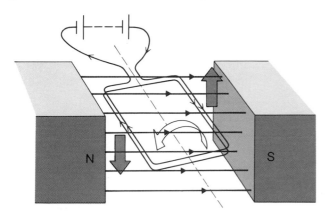

The turning effect is stronger if:
- the current is increased;
- a stronger magnet is used;
- there are more turns on the coil.

Using magnetic forces

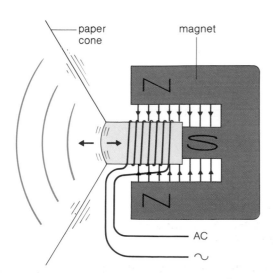

Alternating current passes through the coil of a loudspeaker. The wire in the coil is at right angles to a magnetic field. As the current flows backwards and forwards, the coil is pushed in and out. This makes the cone vibrate and give out sound waves.

1 What is the direction of the force on this wire?

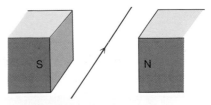

What would the effect on the force of:
a using a higher current?
b reversing the direction of the current?
2 Copy the diagram.

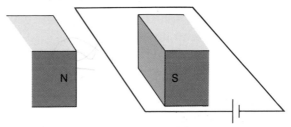

Mark on:
a The direction of the current (conventional).
b The direction of the magnetic field.
c The direction in which the wire will move.
3 Leela has made a battery tester. It uses a magnet, flexible wire and a pointer. With it, she can check whether a small battery is 'live' or 'dead', and find out which terminal is which.

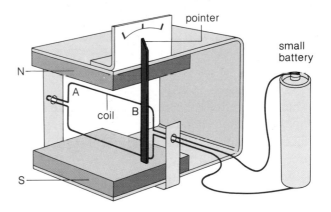

She connects a battery to the tester. The pointer moves to the left.
a Why does the pointer move?
b What is the direction of the magnetic field?
c What is the direction of the current along the top edge of the coil? A-TO-B or B-TO-A?
d Which is the positive (+) end of the battery? TOP or BOTTOM?
e Leela wants to make the tester more sensitive – she wants the pointer to move further when a battery is connected. How could she change her design to make this happen?

4.18 Electric motors

Electric motors use the magnetic turning effect on a coil. They can power anything from a model car to a submarine.

The poles of the **magnet** face one another.

The **coil** is free to rotate between the poles of the magnet.

The **commutator** or split ring is fixed to the coil, and turns round with it.

The **brushes** are two carbon contacts. They connect the coil to the battery.

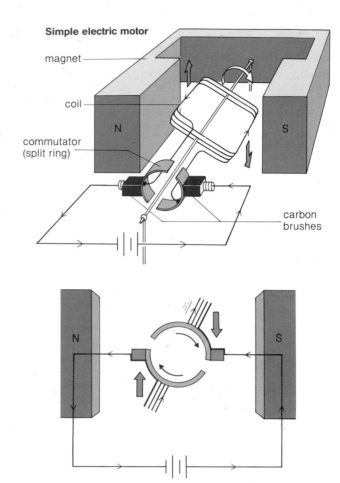

Simple electric motor

magnet — coil — commutator (split ring) — carbon brushes — N — S

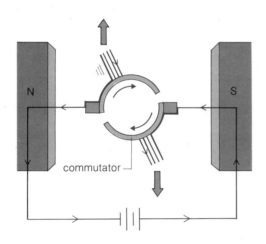

commutator

When a current flows through the coil, one side is pushed up, the other side is pushed down. So the coil turns. However, when the coil is vertical, the forces can't turn it any further because they are pointing the wrong way...

... As the coil shoots past the vertical, the commutator *changes* the current direction. Now the forces point the other way. So the coil is pushed round another half turn. And so on.

Practical motors usually have several coils set at different angles. This gives smoother running and a greater turning effect.

Some motors use electromagnets rather than permanent magnets. This means that they can run on alternating current (AC). As the current flows backwards and forwards through the coil, the magnetic field changes direction to match it. So the turning effect is always the same way.

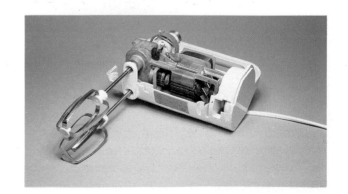

Making a motor

Using a motor kit, you need:

3 metres of plastic-covered copper wire (26 SWG)
1 wooden block with metal tube through centre

1 wooden base	1 iron yoke	2 magnets
1 metal spindle	2 split pins	4 studs
2 rubber rings	3 V battery	Sellotape

1 Insulate one end of the tube with Sellotape.

2 Wind coil on wooden block. You need about 10 turns.

3 Strip plastic from ends of wire.
Fix bare ends of wire to tube using rubber rings.

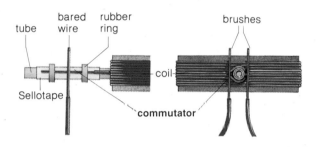

Check the ends are opposite each other and in line with the coil. You have now made the **commutator**.

4 Cut two half-metre lengths of wire.
Bare the ends of the wires.
Fix wires to wooden base using studs.
You have now made the **brushes**.

5 Put split pins into base.

6 Push vertical wires (the brushes) towards each other. Move tube upwards to separate them.

Slide spindle through split pins and tube.
You should now be able to spin the coil.
The brushes should press firmly against the tube and the wire from the coil.

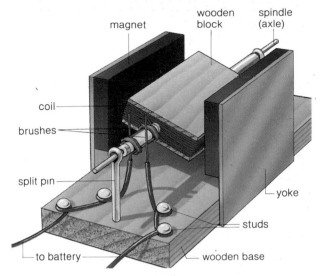

7 Put the two magnets on the yoke to make a single U-shaped magnet. Check that the opposite faces attract each other.

8 Slide yoke into position. Connect wires to battery.
Give the coil a flick to start it turning.

To answer these questions, you will need information from the previous section.

1 COMMUTATOR BRUSHES COIL MAGNET SPINDLE

Which of these:
a is often made from carbon;
b is also known as a split ring;
c turns when current flows through it;
d connects the battery to the split ring and coil;
e changes the current direction every half turn?
2 Someone builds a simple motor following the instructions above. What changes could they make to give the motor a greater turning effect?

3 This is the end view of a simple motor.

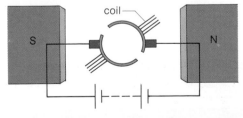

a Copy the diagram and mark on the current direction.
b Which way will the motor turn?
c What is the position of the coil when there is no turning effect on it?
d What is the position of the coil when the current changes direction?

4.19 Electricity from magnetism

You don't need batteries to produce a current. Just a wire, a magnet, and movement.

Moving magnets . . .

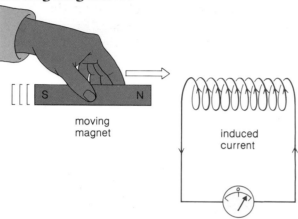

moving magnet

induced current

Push a magnet into a coil. Result – electrons in the coil are given a push by the magnetic field. In other words, a voltage is **induced** or **generated** in the coil. It makes a current flow round the circuit.

For a higher voltage (and higher current):

- move the magnet faster;
- use a stronger magnet;
- put more turns on the coil.

When the magnet stops moving, there is no voltage and no current.

. . . and moving wires

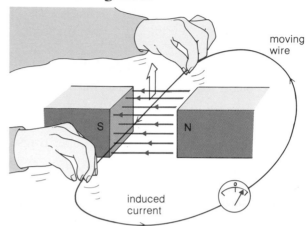

moving wire

induced current

Move a wire across a magnetic field at right angles.

Results – a voltage is induced in the wire. So a current flows round the circuit.

For a higher voltage (and higher current):

- move the wire faster;
- use a stronger magnet.

When the wire stops moving, there is no voltage and no current. If the wire is moved sideways there is no voltage and no current.

> Whenever a conductor cuts through magnetic field lines, a voltage is generated. It doesn't matter whether the conductor moves or the magnet.
> **The faster the field lines are cut, the greater the voltage.**
> This is **Faraday's law**.
> If no field lines are cut there is no voltage.

Playback on a cassette recorder.
Magnetized tape moves over a tiny coil.
A small current is generated.

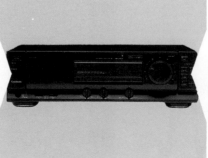

The current is made bigger by an amplifier.
It makes loudspeakers give out sound waves.

Guitar pickups are rows of magnets with coils round them. The steel springs become magnetized. When vibrated, they generate current in the coils.

Current in a wire – which way?

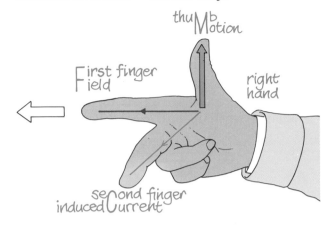

To find which way the induced current flows in a moving wire, use **Fleming's right hand rule**. Hold the thumb and first two fingers of your right hand at right angles. Point them like this, and the second finger gives you the current direction.

Current in a coil – which way?

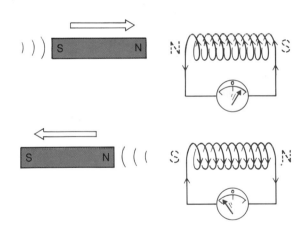

Induced current flows in a coil:
– one way when the magnet is pushed in;
– the opposite way when the magnet is pulled out.
Either way, the coil becomes an electromagnet.
It *repels* the magnet when this is pushed *in*.
It *attracts* the magnet when this is pulled *out*.
Whichever way you move the magnet, you have to move against a force – you have put energy in to get electrical energy out.

This is an example of **Lenz's law**:

an induced current always flows to try and stop the movement which started it.

For a wire at right angles to a magnetic field:

If *current is flowing*, Fleming's *left hand* rule tells you the direction of the force on the wire (see page 182).

If *the wire is moving*, Fleming's *right hand* rule tells you the direction of the induced current (see left).

1 A wire is moved downwards through a magnetic field.

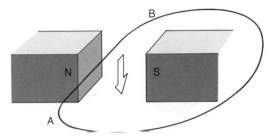

What is the direction of the induced current? A-TO-B or B-TO-A?
What would be the effect of:
a using a stronger magnet;
b moving the wire faster;
c moving the wire upwards;
d moving the wire towards one of the poles?
2 The magnet is being pushed towards the coil.

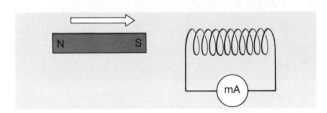

a Copy the diagram. Mark on the type of pole produced at each end of the coil.
b Use the right-hand grip rule (page 178) to work out which way the induced current flows. Mark the direction on your diagram.
c Copy and complete this table to show what happens to the meter when the magnet is moved in and out of the coil:

Magnet pushed in	Needle moves to right
Magnet in coil, but not moving	
Magnet pulled out	
Magnet pushed in again, but faster	

4.20 Generators

Turn a **generator** and out comes a current ...

... current for the lights on your bike.

... current for the circuits in a car.

... current to light up a whole city.

In fact, generators provide over 99% of our electrical energy.

A simple alternator

Most generators give out alternating current (AC).
AC generators are called **alternators**.
This one is providing the current to light up a bulb:

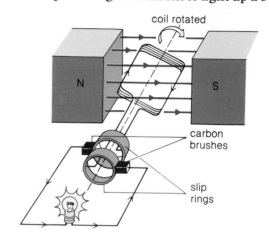

The **coil** is rotated between the poles of a magnet.

The **slip rings** are fixed to the coil and turn round with it.

The **brushes** are two carbon contacts. They rub against the slip rings, so that the rotating coil is connected to the bulb.

When the coil is rotated, it cuts through magnetic field lines. A voltage is generated. This makes a current flow through the bulb.

As the coil rotates, each side travels upwards through the magnetic field, then downwards. So the induced current flows one way, then the opposite way. The current is alternating.

Turning a motor into a generator

You may have made an electric motor like this.
It can be used to generate a current.
Connect the leads to a milliammeter.
Give the coil a spin. The coil cuts through magnetic field lines, and a current is generated.
The milliammeter reading shows that the current is 'one-way' direct current (DC).
But the flow is very uneven.

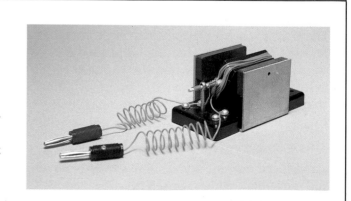

This graph shows how the current from the alternator changes as the coil rotates.

'Forwards' current is plus (+);

'Backwards' current is minus (−).

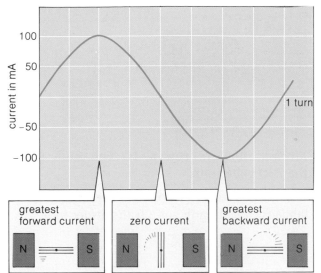

The current is *greatest* when the coil is *horizontal*. The coil cuts field lines most rapidly in this position.

The current is *zero* when the coil is *vertical*. The coil doesn't cut field lines in this position.

The alternator would generate a higher current if:

- the coil had more turns;
- a stronger magnet was used;
- the coil was rotated faster.

Alternator facts

- The 'dynamo' on a bike is an alternator. It produces more voltage the faster you pedal. But you would have to travel at over 50 mph to 'blow' all the bulbs.
- Many alternators use electromagnets instead of permanent magnets. These give a stronger field.
- The alternator in a car is fitted with a 'diode block'. This changes the alternating current into 'one way' direct current. The current supplies the car circuits and keeps the battery charged.
- Alternators in power-stations have to run at a constant speed. If their speed changed, the mains frequency would change from 50 Hz.

To answer these questions, you will need information from the previous section.

1 Copy the boxes. Fill in the first letter of each answer to make a word. This is an essential part of any generator.

?	It measures small currents.
?	Type of current from an alternator.
?	When turned, they produce currents.
?	Field lines leave this pole.
?	Type of energy from a generator.
?	For more current, a generator coil needs more of these.

2 This is the end view of a simple alternator.

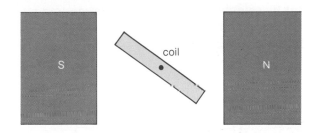

Redraw the diagram to show the position of the coil when the current is **a** greatest **b** zero.

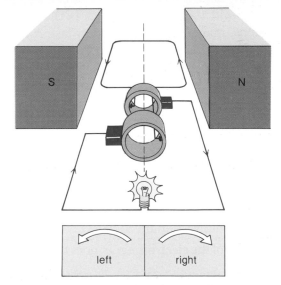

3 This alternator is generating a current.

a Which way is the coil being rotated? To the LEFT? Or to the RIGHT? Use Fleming's right hand rule to find out.

b What changes could you make so that the alternator produced a larger current?

c Explain why the current being generated is AC and not DC.

4.21 Transformers

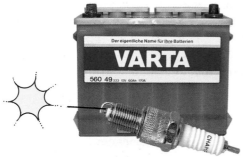

A 12 000 volt spark from a 12 volt battery.
It's needed to ignite the petrol in a car engine. ·
The high voltage is generated in a coil.
Not by pushing a magnet in and out.
But by switching an electromagnet on and off.

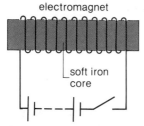
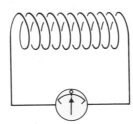

An electromagnet is close to a coil.
Switching on the electromagnet puts a magnetic field around the coil. The effect is the same as pushing a magnet into the coil very fast. A voltage is induced, a current flows, and the meter needed flicks one way. But only for a fraction of a second. When the field is steady, the current stops.

When the electromagnet is switched off, the magnetic field vanishes. The effect is the same as pulling a magnet out of the coil very fast. Just for a moment, the meter needle flicks the other way.

Without the core, the induced voltage would be much less. Can you explain why?

A higher voltage is induced if:

● the core of the electromagnet goes right into the coil;

● the coil has more turns.

The coil in a car engine has many thousands of turns. This gives the thousands of volts needed for the spark plugs. The electromagnet is inside the coil. It runs on only 12 volts. It is switched on and off by a transistor (see page 218) or, in older cars, a set of contacts called 'points'.

Transforming the mains

Problem: how to run a 10 volt bulb from the 230 volt mains.

Answer: use a **transformer.** Easier still – use a labpack. It's already got a transformer in it.

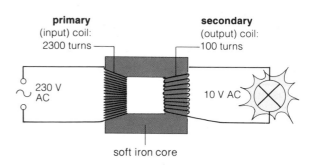

This is one type of transformer.
Alternating current flows through the **primary** or **input coil**. This sets up a changing magnetic field in the **secondary** or **output coil**. The effect is the same as moving a magnet in and out of the coil very fast. An alternating voltage is induced. Alternating current flows. The bulb lights up.

There is a connection between the input and output voltages:

$$\frac{\textbf{input voltage}}{\textbf{output voltage}} = \frac{\textbf{turns on input coil}}{\textbf{turns on output coil}}$$

In this example,

$$\frac{230}{10} = \frac{2300}{100}$$

The input coil has 20 times the turns of the output coil. The **turns ratio** is 23:1.
The input voltage is 23 times the output voltage.

The transformer wouldn't work on DC. Direct current would give a steady magnetic field. So no voltage would be induced in the output coil.

Stepping down and up

There are two types of transformer:

Step-down transformer	Step-up transformer
Symbol:	Symbol:
Turns on output coil less than turns on input coil	Turns on output coil more then turns on input coil
Output voltage less than input voltage	Output voltage more than input voltage
Used in ...	Used in ...
a mains 'power pack' to supply 9 V for a radio or microcomputer	a television to supply 25 000 V for the picture tube

Transformer power

If a transformer doesn't waste any energy, then it must give out as much energy every second as is put in. In other words – its power output must be the same as its power input. For this to happen, the current must change as well as the voltage. For example:

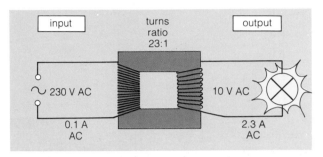

The voltage goes down from:

230 V .... to 10 V

The current goes up from:

0.1 A .... to 2.3 A

The power input
= voltage × current

= 230 × 0.1
= 23 watts

The power output
= voltage × current

= 10 × 2.3
= 23 watts

The voltage has gone *down*. The current has gone *up*. But the power has stayed the *same*.

For any transformer that doesn't waste power:

input voltage × input current = output voltage × output current

1

Transformer	A	B	C	D
Input voltage in V	240	120	50	100
Input turns	1000	1000	1000	2000
Output turns	500	100	2000	2000

Which transformer:
a is a step-up transformer;
b has the same output voltage as input voltage;
c has a turns ratio of 10:1;
d has an output voltage of 12 V;
e has the highest output voltage?

2 A 23 V bulb takes a current of 2 A. Its power supply is a transformer connected to the 230 V mains.

You have to choose a suitable transformer. You can assume that the transformer wastes no power.

a What turns ratio is needed?
b What power is taken by the bulb?
c What power is taken from the mains?
d What current is taken from the mains?
e Why wouldn't the transformer work on DC?

4.22 Power across the country

Mains power comes from huge alternators in power-stations.

Transformers step up the voltage before the power is carried across country by overhead cables.

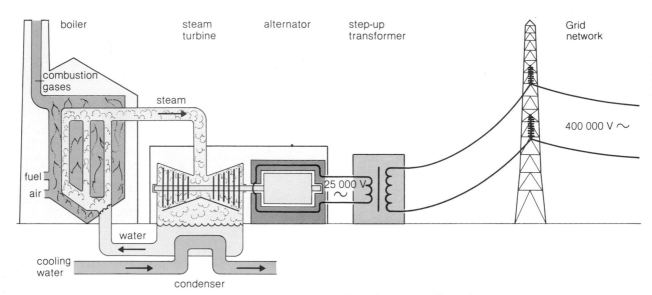

boiler | steam turbine | alternator | step-up transformer | Grid network

combustion gases

steam

fuel
air

25 000 V ~

400 000 V ~

water

cooling water

condenser

The power-station

In the power-station, the alternators are driven by huge **turbines**, spun round by the force of high-pressure steam. The steam is made by heating water in a boiler. The heat comes from burning coal, gas or oil – or from a nuclear reactor (see page 209).

Huge **cooling condensers** change the steam back into water for the boiler. The condensers need vast amounts of cooling water. So power-stations are often built near rivers or the sea.

Why change voltage?

The current from a large alternator can be 20 000 amperes or more. It needs very thick, heavy and expensive cable to carry it. So a transformer is used to *step up* the voltage and *reduce* the current. Then thinner, lighter and cheaper cables can be used to carry the power across country.

If a 25 000 V alternator produces a current of 20 000 A, its power output is 500 000 000 W.

If the voltage is stepped up to 400 000 V, the current *drops* to 1250 A but the power is still 500 000 000 W.

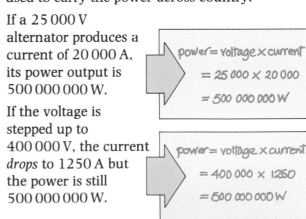

power = voltage × current
= 25 000 × 20 000
= 500 000 000 W

power = voltage × current
= 400 000 × 1250
= 500 000 000 W

Why AC?

If power-stations didn't generate AC, transformers couldn't be used to change the voltage. DC voltages can be changed, but this is difficult and expensive. Transformers don't work on DC.

Then transformers reduce the voltage ...

...before the power is supplied to homes, offices and factories.

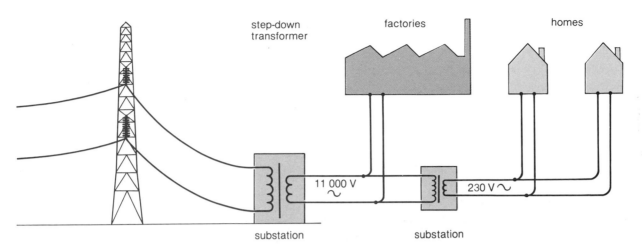

step-down transformer

factories

homes

11 000 V

230 V

substation

substation

The Grid

Each power-station feeds its power to a network of cables and switching stations called the **Grid**. If one area of the country needs more power, it can be supplied by power-stations in other areas.

To answer these questions, you may need information from the previous section.

1 MW = 1 000 000 W

1 Explain why:
a steam is needed in a power-station;
b power-stations are often near a river or the sea;
c the voltage is stepped up before power is fed to the overhead cables;
d power-stations generate AC, and not DC.
2 The three power-stations were built to supply the towns of Newleigh, Extown, and Oldwich.

Each alternator generates 20 000 V.
Each has a power output of 100 MW.
The stations feed their power to the Grid.
The towns take their power from the Grid.

a What is the Grid?
b How many alternators must be working to supply the town of Newleigh?
c Which towns could be supplied by A station alone?
d How much spare power is being supplied to the Grid for use in other areas?

Power from the Grid

Power from the overhead cables is fed to substations. Here, the voltage is stepped down by transformers. Homes take their power at 230 volts. Factories and hospitals take their power at higher voltages.

e If B station shuts down, how much power must be supplied to the towns from other parts of the country?
f How much energy (in joules) does each alternator supply in 1 second?
g How much current is being generated by each alternator?
h After the voltage has been stepped up to 400 000 V, how much current does each alternator supply to the Grid?

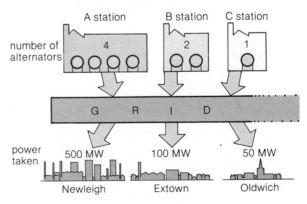

A station
B station
C station

number of alternators

4
2
1

G R I D

power taken

500 MW
100 MW
50 MW

Newleigh
Extown
Oldwich

4.23 Power plus

Hidden power

In Snowdonia, the landscape hasn't been spoilt by pylons, because the electricity company has put its power lines underground. But this is an expensive way of sending power. Lower voltages have to be used, and that means higher currents and thicker cables. To save the landscape, people have to pay more for their electricity.

Guaranteed power

Power cuts don't happen very often. But when they do, the results can be serious:
A loss of power here – and 200 cows will need milking by hand.

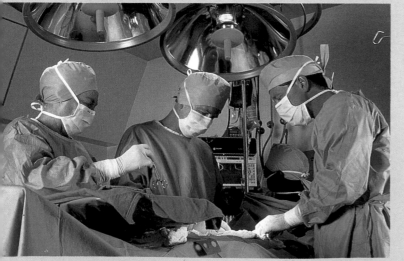

A loss of power here might put someone's life at risk.

For emergencies, most large hospitals and farms have stand-by generators. They are driven by engines which run on petrol, diesel or bottled gas. They start up automatically if there is a mains failure.

Extra power

Battery hens? In Somerset, hens have solved one farmer's electricity supply problems. He saves their droppings in a tank and collects the gas given off. Then he uses the gas to run the engine which drives his generator.

In Florida, the police have collected so much marijuana in drugs raids, that a power-station has been specially converted to burn it. One tonne of marijuana gives nearly as much heat as three barrels of oil.

In Edmonton, North London, the council has turned one of its rubbish incinerators into a generating station. Electricity is generated using the heat from burning household rubbish. The council gets rid of its waste. And it keeps its costs down by selling the electricity to the local electricity company.

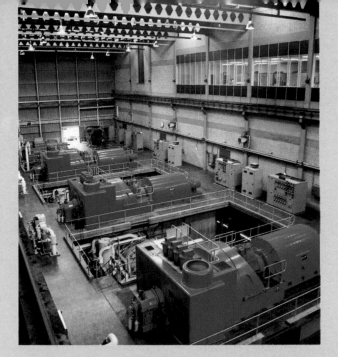

Acid rain

All over Europe, acid rain is falling. The acid in the rain is only weak. But it is killing fish in the lakes of Norway and Sweden, damaging forests in Germany, and eating into the stonework of old buildings.

In Norway, Sweden and Germany, they blame the sulphur fumes from Britain's coal-burning power-stations. Winds are blowing the fumes across Northern Europe. And the fumes are making the rain acid.

The electricity companies don't agree. They say that there is no firm evidence to link their power-stations with acid rain. Sulphur fumes aren't a new problem – factories and road vehicles have been producing them for years.

And as the argument goes on, the acid rain still falls.

Water power

Hydroelectric power. A river is dammed to form a lake. Water rushes from the lake to turn generators at the foot of the dam. No pollution. But the landscape is changed, and local animal and plant life is disturbed.

In Sweden, much of their electricity comes from hydroelectric schemes. But plans to build more dams have been dropped. The Swedes don't want to see more countryside destroyed.

In Sri Lanka, they are keen to expand their use of hydroelectricity. Unless they can supply more power from their own resources, they will have to borrow more money to buy oil. And that will keep the country poor.

Make a list of the buildings where you think an emergency generator is essential.

Here are some of the ways in which a town could get its power:

a hydroelectric power scheme;
a coal-burning power-station;
a small generator in every building.

How many advantages and disadvantages can you think of for each?

Try to find out:
– where the power-stations in your region are sited;
– what fuels they use.

Questions on Section 4

1 The circuit below can be used to test whether different materials allow an electric current to pass through or not:

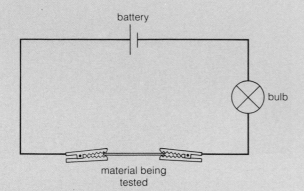

battery

bulb

material being tested

a Will the bulb light up when each of the following materials is connected between the clips? (Answer each with a YES or NO.)
  i Iron
  ii Plastic
  iii Glass
  iv Copper
  v Aluminium
b What name is given to a material that:
  i lets an electric current pass through?
  ii stops an electric current passing through?

2 Use your ideas about electrons to explain each of the following:
  a Electrons are pushed out of the negative ($-$) terminal of a battery, not the positive ($+$).
  b If you rub a polythene rod with a piece of cloth, the rod ends up with negative charge ($-$), while the cloth is left with an equal amount of positive charge ($+$).
  c Unlike most other materials, metals are good conductors of electricity.

3 These wires are made of the same material. They all have resistance.

A

B

C

a Which wire has most resistance?
b Which wire has least resistance?

4 a Below, ammeters are being used to measure the current in different parts of a circuit.

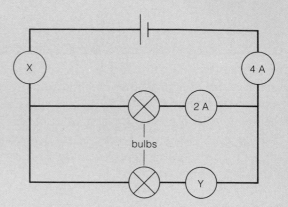

X

4 A

2 A

bulbs

Y

  i Have the bulbs been connected to the battery in *series* or in *parallel*?
  ii What is the reading on ammeter X? Give a reason for your answer.
  iii What is the reading on ammeter Y? Give a reason for your answer.
  iv If one of the bulbs is removed, so that there is a gap between its connecting wires, what will be the effect on the other bulb?
b Below, voltmeters are being used to measure the voltage across different parts of a circuit.

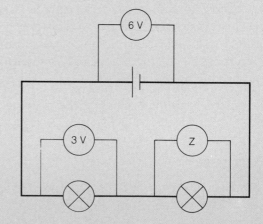

6 V

3 V

Z

  i Have the bulbs been connected to the battery in *series* or in *parallel*?
  ii What is the reading on voltmeter Z? Give a reason for your answer.
  iii If one of the bulbs is removed, so that there is a gap between its connecting wires, what will be the effect on the other bulb?

5 In the circuit below, a 6 V battery has been connected to a 3 Ω resistor:

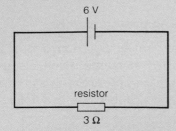

a What is the current in the circuit, in A?
b What is the power output of the battery, in W?
c How much energy, in joules, does the battery give to each coulomb of charge pushed out?
d How much charge, in coulombs, flows through the resistor every second?

6 The fan heater below runs on the 230 V mains. It has two heating elements, each taking a current of 4 A, and a fan taking a current of 0.5 A.

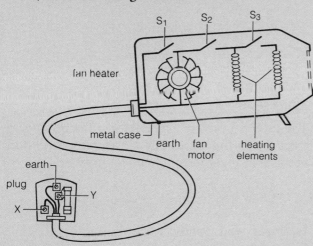

a Which switch should be closed if you want:
  i the fan only, for cool air?
  ii the fan and one element?
b If the fan heater is on full power, how much current flows from the mains?
c What fuse should be fitted in the plug, 3 A or 13 A?
d Which terminal in the plug is live?
e Why is the earth connection needed?
f What colour cable should be connected to terminal X?
g What colour cable should be connected to terminal Y?
h Give reasons for each of the following:
  i If an appliance takes a current of 2 A, you should not fit a 13 A fuse in its plug.
  ii Some appliances, such as radios and hair-dryers, do not need an earth wire.

7 A low-energy light bulb has a power rating of only 20 W, but produces the same amount of light as an ordinary 100 W bulb. Both bulbs are to be switched on for 1000 hours.

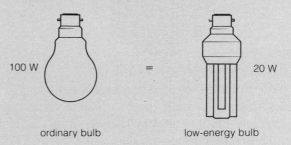

a What is the power of each bulb in kW?
b How much energy (in kW h) will the low-energy bulb use in 1000 hours?
c How much energy (in kW h) will the ordinary energy bulb use in 1000 hours?
d If electricity costs 10p per unit (kW h), how much money will be saved on electricity bills over a 1000-hour period by using the low-energy bulb instead of the ordinary bulb?
e The ordinary bulb uses more energy than the low-energy one for the same amount of light. What happens to the extra energy supplied?

8 Here is an experiment involving two circuits.

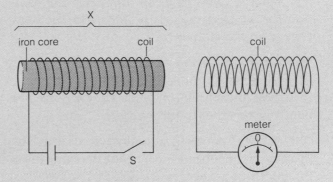

a What name is given to the device marked X?
b Describe what you will observe if you are looking at the meter:
  i at the moment switch S is closed;
  ii a few seconds later, if switch S is kept closed;
  iii at the moment switch S is opened again.
c Name a device which makes use of the principle shown in the experiment above.
d Describe another way in which you could make a current flow in the right-hand circuit without connecting anything else into the circuit.
e With switch S, the magnetic field from the core can be turned on and off. What difference would it make if the core were made of steel rather than iron?

5.1 Inside atoms

60 000 000 000 000 000 000 000

That's the number of atoms in this penny. Everything is made from atoms. They are extremely small. And they're made from particles which are very much smaller.

This is one simple picture or **model** of the atom: At the centre, is a tiny **nucleus** made up of particles called **protons** and **neutrons**. Around this, orbit much smaller particles called **electrons**.

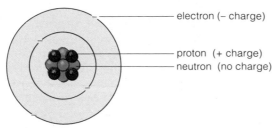

The charges on electrons and protons are equal, but opposite. Normally, an atom has the same number of electrons as protons. So the total charge on the atom is zero.

The force of attraction between + and − charges hold the electrons in orbit around the nucleus. A **strong nuclear force** binds the protons and neutrons together in this nucleus.

The mass of an atom is mainly in its nucleus. Protons and neutrons have about the same mass. But electrons are very much lighter. It takes about 2000 electrons to make up the mass of a proton.

The nucleus is far too small to show in a diagram. If an atom were the size of this concert hall, its nucleus would be smaller than a pea.

Elements, and proton number

All things are made up from about 100 basic materials called **elements**. Each element has a different number of protons in its atoms. It has a different **proton number**. For example:

Element	Chemical symbol	Proton number
Hydrogen	H	1
Helium	He	2
Lithium	Li	3
Beryllium	Be	4
Boron	B	5
Carbon	C	6
Nitrogen	N	7
Oxygen	O	8
Radium	Ra	88
Thorium	Th	90
Uranium	U	92
Plutonium	P	94

The proton number also tells you the number of electrons in the atom. The electrons control how atoms join together in chemical reactions.

Isotopes, and nucleon number

The atoms of an element aren't all alike. Some have more neutrons than others. These different 'versions' of the atom are called **isotopes**. Most elements are a mixture of two or more isotopes. You can see some examples in the chart on the opposite page.

For example:

The metal lithium is a mixture of two isotopes. These are lithium-6 and lithium-7. Lithium-7 is the more common. Over 93% of all lithium is lithium-7.

The total number of protons and neutrons in each atom is called the **nucleon number**. Isotopes have *different* nucleon numbers. But they have the *same* proton number.

Some atoms, elements, and isotopes

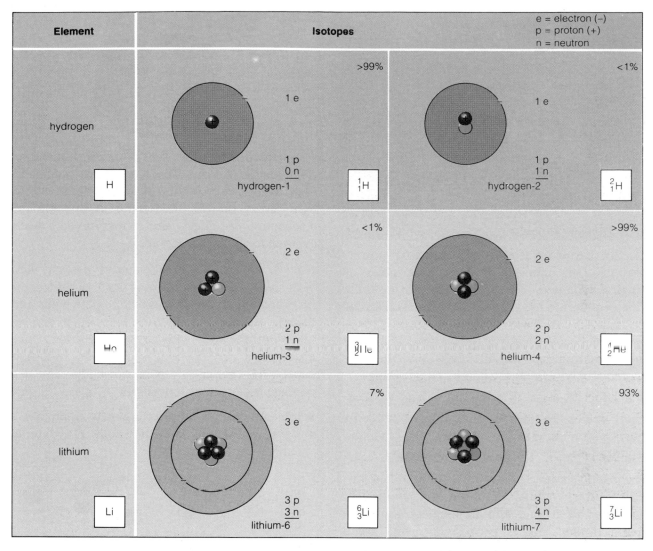

	electrons	protons	neutrons	nucleon number
sodium-23	11			
aluminium-27	13			
strontium-90	38			
cobalt-60	27			

1

electron	proton	neutron	nucleus

Which of these:

a orbits the nucleus;

b is a particle with a + charge;

c is uncharged;

d is lighter than all the others;

e is made up of protons and neutrons;

f has a – charge?

2 Copy the chart and fill in the blanks:

To answer the following questions, you will need information from the table of elements on the opposite page.

3 Nitrogen-14 can be written $^{14}_{7}N$. How can the following be written?

a Radium-226; **b** Uranium-235;

c Oxygen-16; **d** Carbon-12.

4 Here is some information about four atoms:

Atom A: 3 electrons, nucleon number 7;

Atom B: 142 neutrons, nucleon number 232;

Atom C: 3 neutrons, nucleon number 6;

Atom D: 5 electrons, nucleon number 11.

What elements are A, B, C and D?

Which pair of atoms are isotopes?

5.2 Nuclear radiation

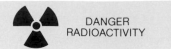

Deadly cargo. Waste from a nuclear power-station. The material in each flask gives off **nuclear radiation** – it is **radioactive**. The flasks are thick enough to absorb the radiation. And strong enough to withstand a head-on crash. This is essential. Because if any of the waste leaked out, it could contaminate the air, crops, and the local water supply. And many deaths could result.

Radioactive atoms are unstable atoms. In time, each nucleus breaks up and shoots out a tiny particle or a burst of wave energy. This 'radiates' from the nucleus – it is nuclear radiation.

Radioactive materials aren't only found in nuclear power-stations. There are tiny amounts of them in the ground, the atmosphere, and even living things. This is because elements are a mixture of isotopes, and some isotopes are unstable.

For example:

Isotopes		
Stable	Unstable, radioactive	Found in
carbon-12 carbon-13	carbon-14	air, plants animals
potassium-39 potassium-41	potassium-40	rocks, plants sea water
	uranium-234 uranium-235 uranium-238	rocks

Danger! Ions

People who work in nuclear power-stations wear a film badge like this. This reacts to nuclear radiation rather like the film in a camera reacts to light. Every month, they hand in the film for developing, to check that they haven't been exposed to too much radiation.

Nuclear radiation can damage or destroy vital body cells. An overdose may lead to cancer or incurable radiation sickness. Radioactive gas and dust is especially dangerous because it can be taken into the body with air, food, or drink. Once absorbed, it can't be removed, and its radiation causes cell damage deep in the body. The cell damage is caused by **ionization**:

- Ions are charged atoms (or groups of atoms).
- If a material is **ionized**, then some of its atoms carry + or − charges.

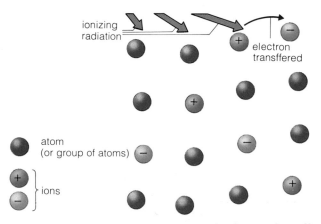

- Nuclear radiation, X-rays, and ultraviolet all cause ionization. When they strike atoms, they wrench electrons from them, leaving them +. The electrons are caught by other atoms, making them −.

- Living cells are very sensitive to ionization. It can completely upset their life processes.

There are several types of nuclear radiation. In a nuclear power-station, the radiation from the nuclear fuel is mainly a stream of neutrons (see page 208). The radiation from the radioactive waste is mainly **alpha**, **beta** and **gamma** radiation:

type	alpha particles (α)	beta particles (β)	gamma rays (γ)
	each particle is 2 protons + 2 neutrons	each particle is an electron	electromagnetic waves similar to X-rays
charge	**+**	**−**	no charge
mass	heavy compared with betas	very light	. . .
speed	up to $\frac{1}{10}$ speed of light	up to $\frac{9}{10}$ speed of light	speed of light
ionizing effect	strong	weak	very weak
penetrating effect	not very penetrating: stopped by thick sheet of paper, or skin	penetrating: stopped by about 5 mm of aluminium	highly penetrating: never fully absorbed but 25 mm of lead halves strength
effect of fields	bent by magnetic and electric fields	bent strongly by magnetic and electric fields	not bent by magnetic or electric fields

The diagram below shows how alpha, beta, and gamma radiation are affected by a strong magnetic field. The alpha beam is a flow of positively (+) charged particles, so it is equivalent to an electric current. The beam is forced upwards, as predicted by Fleming's left hand rule (see page 182). The beta particles are much lighter than the alphas and have a negative (−) charge, so they are bent more and in the opposite direction. Gamma rays are not bent by the magnetic field because they are uncharged.

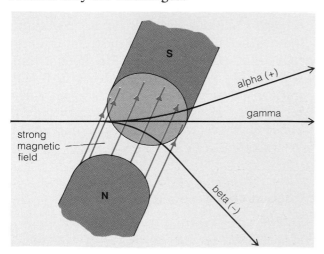

1 Copy the boxes. Then use the first letter of each answer to form a word; clue: this is where nuclear radiation comes from.

?	Uncharged parts of an alpha particle.
?	If a material is radioactive, its atoms are . . .
?	Radioactive material in air, plants and animals.
?	Metal used for absorbing gamma rays.
?	Beta particles.
?	Material with nucleon number of 235.
?	If a material isn't radioactive, its atoms are . . .

2

alpha	beta	gamma

Which type of radiation:
a is uncharged;
b can penetrate lead;
c is stopped by a thick sheet of paper;
d has a + charge;
e has most mass;
f isn't bent by electric or magnetic fields;
g is made up of electrons?

3 Explain:
a how an ionized material is different from one that isn't ionized;
b why nuclear radiation is harmful to living things.

5.3 Living with radiation

Nuclear accidents are rare compared with other types of accident – such as air crashes, fires, or dam collapses.

More nuclear power-stations are essential because the world's supplies of oil, coal, and natural gas are running out.

Britain's worst nuclear accident

Nuclear power – how safe?

The disaster at Chernobyl nuclear power-station, Ukraine, happened swiftly, without warning. It was in the early hours of April 26 1986 when the cooling system of number four reactor failed. Minutes later, a violent explosion blew the top off the reactor and blasted a huge cloud of radioactive gas high into the atmosphere. Two people were killed outright. Hundreds received massive radiation overdoses. And more than 25 000 had to be evacuated from their homes. Days later, the radioactive cloud had spread as far as Scotland. Its radiation was weak but, all over Europe, radioactive rain was falling. In some areas, people were advised not to eat fresh vegetables, or drink fresh milk, and the sale of meat was banned.

The accident at Chernobyl was the world's worst nuclear accident. In Britain, it convinced many people that all nuclear power-stations should be shut down. But the company running the power-stations doesn't agree. It claims that:

- A similar disaster can't happen in Britain, because the reactors are of a much safer design.
- Fewer deaths are caused by the use of nuclear fuel than by mining for coal or drilling for oil and gas.

Windscale (now called Sellafield) in Cumbria. Here, in 1957, a nuclear reactor overheated and caught fire. No one was killed outright, but fourteen workers received radiation overdoses. And small amounts of radioactive gas and dust were released over the local countryside.

An official report said that accident had nearly become a full-scale disaster. The nuclear authorities wanted the report published. But the Prime Minister at the time refused. He thought that it would make people lose confidence in Britain's nuclear industry. Thirty years later, the cabinet records for 1957 were released. Only then did the public discover what had really happened at Windscale.

The search for a nuclear dustbin

To the nuclear authorities, the village of Elstow, near Bedford, seemed like the ideal dumping site for Britain's nuclear waste. Firm rocks underneath meant that the containers wouldn't crack open. And the soft clay soil above would absorb the radiation.

When the dumping plans were announced, the residents of Elstow reacted angrily. There was a storm of protests and the plans were dropped. Now, the authorities are searching for other sites. They've considered drilling storage tunnels under the North Sea. But Norway and Denmark are fiercely opposed to this idea. They're afraid that radioactive materials might leak out and contaminate their coastlines. Meanwhile, the waste from Britain's nuclear power-stations is piling up. It's going to be radioactive for hundreds of years. And it's got to be stored somewhere . . .

Gammas keep fruit fresher

These strawberries were picked three weeks before the photograph was taken.

So were these. But these were put straight into a beam of gamma radiation. The radiation stopped the rotting process. So the strawberries look as fresh as the day they were picked. They haven't become radioactive. And their taste has hardly changed at all.

Irradiating food has many advantages. Or so the food producers claim. The radiation stops vegetables sprouting when they are being stored. It kills off the mould which makes food go off. And it destroys bacteria like *Salmonella* which can give you food poisoning. Many supermarkets want to sell fruit treated with radiation. They claim that it will mean better quality for their customers, less waste and lower prices.

But not everyone likes the idea. The radiation may destroy important vitamins. And it may change some of the chemicals in food – so that they behave like dangerous additives. Irradiated food does last longer. But when you next buy fresh strawberries, how fresh will they really be?

Most people wouldn't want a nuclear waste dump near their homes, even if they were told that the dump was completely safe. Why not? How many reasons can you think of?

Some supermarkets want to sell fruit and vegetables that has been treated with radiation. Make lists of the points for and against this scheme.

5.4 Radiation: detecting it . . .

The Geiger-Müller tube

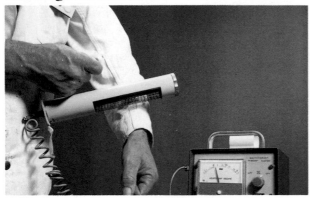

Radiation check on a nuclear laboratory worker. The instrument is a **Geiger-Müller tube**. If there is any radioactive dust on the clothing, the G-M tube will detect the radiation from it.

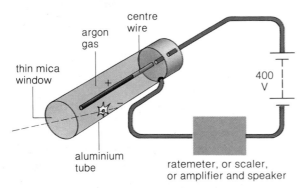

The 'window' at the end of the tube is thin enough even for alpha particles to pass through. If an alpha particle enters the tube, it ionizes the gas inside. This sets off a high-voltage spark across the gas, and a pulse of current flows in the circuit. A beta particle or a burst of gamma radiation would have the same effect.

The G-M tube can be connected to:

A ratemeter. The needle gives a reading in counts per second. If 50 alpha particles are detected by the tube every second, the reading is 50 counts per second.

A scaler. This counts the total number of particles or bursts of gamma radiation entering the tube.

An amplifier and speaker. The speaker gives a 'click' every time a particle or burst of gamma radiation enters the tube.

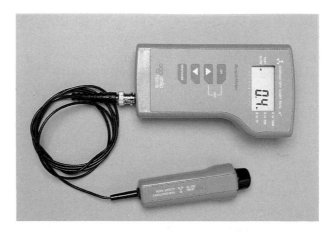

The cloud chamber

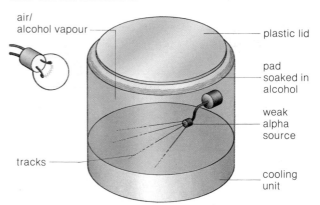

A cloud chamber is particularly useful for studying alpha particles. You can actually see their tracks. The chamber has cold alcohol vapour mixed with the air inside it. The alpha particles make the vapour condense. So you see a trail of tiny droplets where each particle passes through.

Tracks like this show that alpha particles have a range of only a few centimetres in air.

and using it

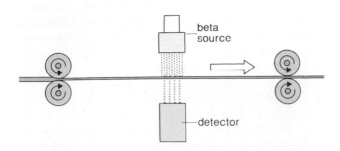

Checking the thickness in a tyre factory.
This moving band of tyre cord has a beta source on one side and a detector on the other. If the detector picks up too much radiation, the cord is being made too thin.

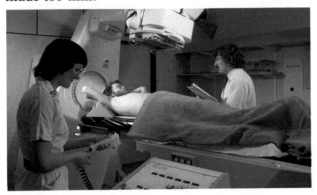

Killing cancer cells.
Gamma rays kill living cells. This machine can concentrate gamma rays on cancer cells in one small area of the body.

To answer these questions you will need some information from the previous Unit.

1 | ratemeter | cloud chamber | scaler | amplifier |

Which if the above would be:
a used to show the tracks of alpha particles;
b connected to a G-M tube to measure the total number of particles received;
c connected to a G-M tube to measure the number of particles received every second.
2 Explain why:
a the window at the end of a G-M tube has to be very thin;
b a cloud chamber has cold alcohol vapour in it;
c a gamma ray source might be put inside a steel pipe.

Seeing through steel.
Gamma rays are used to take X-ray-type photographs of welded metal joints. Here, the gamma source is inside the pipe. The X-ray film is wrapped round the outside of the pipe.

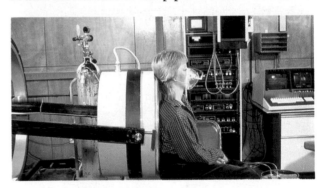

Checking the lungs with a radioactive tracer. Tiny amounts of radioactive krypton gas are breathed in by the patient. The flow of gas can be 'watched' on a TV screen.

3 Tracy carries out experiments with three radioactive sources. She puts different absorbing materials in front of each, and measures the radiation passing through. These are her readings:

| Source | Radiation received: counts per second | | | |
	no absorber	thick cardboard	2 cm of aluminium	2 cm of lead
A	8	8	7	5
B	12	5	4	3
C	10	9	0	0

Tracy knows that: one source gives out beta particles; another source gives out gamma rays; the third source gives out two types of radiation.

Can you work out what is coming from each source?

5.5 Radioactive decay

The breaking up of unstable atoms is called **radioactive decay**. It's happening around you all the time – in the rocks, in the atmosphere, and even in your house. A G-M tube can detect this weak **background radiation**. Each count or 'click' means that one particle or burst of gamma radiation has gone into the tube. Somewhere, one atom has decayed.

Radioactive decay is completely random. You can't tell which atom is going to break up next, or when. But some types of atom are more unstable than others. They decay at a faster rate.

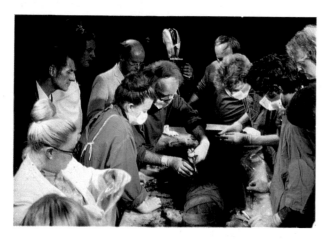

Scientists unwrap an ancient Egyptian mummy. By measuring the radiation it gives off, they can find out its age. Tiny amounts of radioactive carbon-14 pass in and out of all living things but some stays trapped in the body at death. As time goes on, the radiation from the carbon gets weaker and weaker. Scientists can use this fact to work out the age. It's called carbon dating.

Half-life

A G-M tube is placed next to a large sample of radioactive iodine-128. Both are shielded from background radiation. This is how the count-rate changes with time:

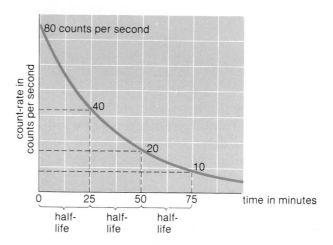

As time goes on, there are fewer and fewer unstable atoms left to decay. So the count-rate gets less and less. After 25 minutes, the count-rate has fallen by half its original value. Half the atoms have now decayed. After another 25 minutes, the count-rate has halved again. And so on. Iodine-128 has a **half-life** of 25 minutes. Here are the half-lives of some other radioactive materials:

Radon-222	4 days
Strontium-90	28 years
Radium-226	1602 years
Carbon-14	5730 years
Plutonium-239	24 400 years
Uranium-235	700 000 000 years

Hot rocks

Deep in a mine, 100 metres underground, the temperature of the rock is 50 °C. 1000 metres further down, it is hot enough to boil water. The heat comes from radioactive atoms in the rocks. When they decay, they shoot out alpha or beta particles (and gamma rays) which hit other atoms and make them move faster. Energy released from the nucleus is changed into heat energy.

Dating rocks

When rocks are formed, natural radioactive isotopes become trapped. For example, in mica rock, over millions of years, the amount of radioactive potassium-40 becomes less, while the amount of argon-40 decay product increases. By measuring the ratio of potassium to argon, the age of the rock can be estimated.

New atoms from old

When an atom decays, it may change into a completely different atom. For example:

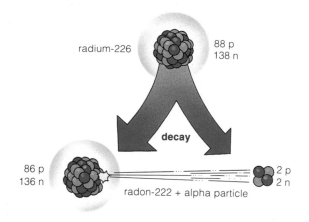

When an atom of radium-226 gives off an alpha particle, it loses 2 neutrons and 2 protons.
It becomes a charged atom of radon-222.
This new atom is a **decay product** of radium-226.
It too is radioactive.

Decay hazards

Sellafield nuclear waste reprocessing plant, Cumbria. Radioactive decay products from Britain's nuclear power-stations are sent here to be separated. Many of the products are highly radioactive. And many have long half-lives. They will need to be stored for hundreds of years before their radiation has dropped to a safe level.

Some decay products are especially dangerous:

Strontium-90 and **iodine-131** are easily absorbed by the body. Strontium becomes concentrated in the bones, iodine in the thyroid gland.

Plutonium-239 is the most dangerous substance of all. Breathed in as dust, the smallest amount can kill.

1 Look at the table of half-lives on the opposite page. If small amounts of strontium-90 and radium-226 both gave the same count-rate today, which would give the higher reading in 10 years' time?
2 A G-M tube is placed near a weak radioactive material in the lab. There is no shielding round the tube or source to absorb background radiation. This is the graph of count-rate against time, using readings taken every half minute:

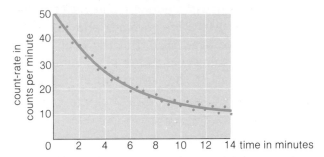

a Why are the points on the graph not on a smooth curve?
b Why does the graph level out above the zero?
c What is the half-life of the material? (Give an approximate value, to the nearest minute.)
3 This is how the count-rate from a radioactive material changed:

Time in s	0	20	40	60	80	100	120
Count-rate in counts/s	57	44	33	25	19	14	11

a Plot a graph of COUNT-RATE (*side axis*) against TIME (*bottom axis*).
b How long did it take the count-rate to fall from 50 to 25 counts/second?
c How long did it take the count-rate to fall from 40 to 20 counts/second?
d What is the half-life of the material?
4 The explosion at the Chernobyl reactor released a cloud of radioactive gas and dust into the atmosphere. It contained caesium-137 (half-life 30 years) and iodine-131.
a Here are some measurements of the count-rate from a small amount of iodine-131:

Time in days	0	4	8	12
Count-rate in counts/s	240	170	120	85

What is the half-life of iodine-131?
b Two months after the explosion, scientists were still concerned about the health risks from the caesium, but felt that the iodine was no longer a threat. Can you explain why?

5.6 Nuclear power

Hinkley Point 'B' nuclear power-station, in Somerset.

It can generate enough power to supply a large city. Like most power-stations, it uses heat to make steam which drives the turbines and turns the generators (see page 192). But the heat doesn't come from burning coal, gas or oil. It comes from uranium atoms as they break up in a **nuclear reactor**.

Fission

Natural uranium is a dense radioactive metal. It is made up mainly of two isotopes: uranium-238 (over 99%), and uranium-235 (less than 1%).

Both uranium isotopes decay naturally. But uranium-235 nuclei can be made to break up very quickly.
They can be split by neutrons.
This is called **fission**.
Fission releases energy much faster than natural decay. And it works like this:

A beam of neutrons is directed at the uranium. If a neutron strikes a nucleus of uranium-235, this splits into two roughly equal parts, and shoots out two or three neutrons as well.
If these neutrons hit other uranium-235 nuclei, they make them split and give out more neutrons. And so on. The result is a **chain reaction**:

When a uranium-235 nucleus splits, energy is released. As the bits are thrown violently apart, they bump into other nuclei and make them move faster. In this way, nuclear energy is changed into heat energy.

If the chain reaction is *uncontrolled*, huge numbers of nuclei are split in a very short time. The heat builds up so rapidly that the material bursts apart in an explosion. This happens in a nuclear bomb.

If the chain reaction is *controlled*, there is a steady output of heat. This happens in a nuclear reactor.

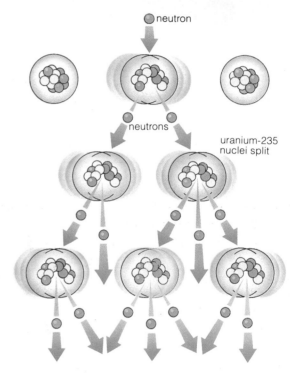

neutron

neutrons

uranium-235 nuclei split

1 Copy the boxes. Fill in the first letter of each answer to make a word. These split atoms in a reactor:

?	Centre of an atom.
?	This is released during fission.
?	Material used for fission.
?	This rises when the control rods are raised.
?	Absorbed by the concrete around a reactor.
?	Non-nuclear fuel used in some power-stations.
?	Type of energy stored in the nucleus.
?	What atoms do during fission.

Nuclear reactors

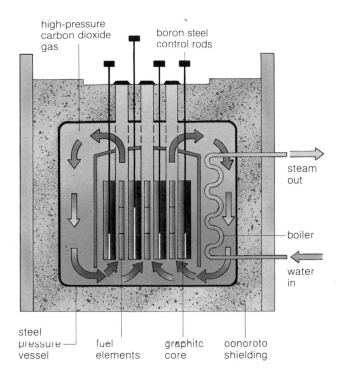

high-pressure carbon dioxide gas

boron steel control rods

steam out

boiler

water in

steel pressure vessel

fuel elements

graphitc core

oonoroto shielding

This is a gas-cooled reactor like the one at Hinkley Point 'B'. The reactor is inside a steel pressure vessel. It is surrounded by thick concrete to absorb radiation. Heat from the reactor is carried away by carbon dioxide gas. The hot gas heats the water in the boiler.

- A reactor can't explode like a nuclear bomb. The uranium-235 atoms are too spaced out for an uncontrolled chain reaction to occur.

2 | boron | graphite | uranium | concrete | steel |

In a reactor, which of these materials is used:
a as a moderator;
b to absorb radiation;
c in the nuclear fuel elements;
d in the control rods?

3 Explain why, in a gas-cooled reactor:
a a moderator is needed;
b carbon dioxide gas is pumped through the reactor;
c the chain reaction stops if the control rods are kept fully lowered.

The reactor contains:

Nuclear fuel elements These contain uranium dioxide. They contain natural uranium with extra uranium-235 mixed in. 1 kg of this 'enriched' uranium fuel gives as much energy as 25 tonnes of coal.

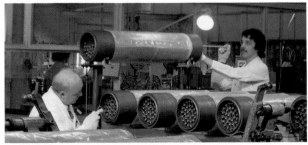

A graphite core This is a **moderator** – it slows down the neutrons released by fission.
The neutrons have to be slowed down otherwise the chain reaction would stop. The uranium-238 atoms in the fuel absorb fast neutrons. But they don't absorb slow ones.

Control rods. These are raised or lowered to control the rate of fission. They are made of boron steel: the boron absorbs neutrons.

If the rods are raised, more neutrons can cause fission. So the reactor temperature rises.
If the rods are fully lowered, the chain reaction stops and the reactor cools down.

- In a pressurized water reactor, water is used to carry heat from the reactor core.

4 Copy and complete the chart to show the energy changes which take place in a nuclear power-station. Each of the question marks is a type of energy.

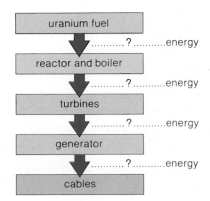

uranium fuel

..........?..........energy

reactor and boiler

..........?..........energy

turbines

..........?..........energy

generator

..........?..........energy

cables

5.7 The oscilloscope

A **cathode ray oscilloscope (CRO)** looks a bit like a TV. But you can use it to study sound waves, AC voltages, and even heartbeats.

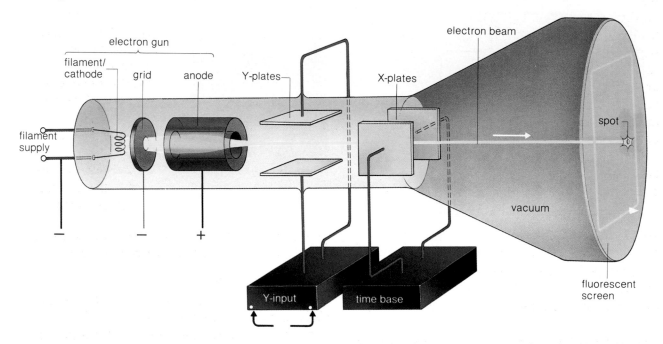

The electron gun makes a narrow beam of electrons. It has a cathode ($-$) to give off electrons, and an anode ($+$) to accelerate them. In the type shown above, the cathode is a tiny tungsten filament, which is heated by a current. The heat makes electrons leave the metal. The effect is called **thermionic emission**.

The Y-plates are used to bend the beam up or down. This is done by putting a voltage across the **Y-input terminals** so that one plate goes $+$ and the other goes $-$. If AC is put across the plates, they go $+$, $-$, $+$, $-$, and so on. The spot moves up and down so fast that it looks like a line.

The grid controls the brightness. When the grid is made negative ($-$), it pushes back some electrons. So fewer of them reach the screen.

The screen is coated in a fluorescent material. You see a bright spot where the electron beam strikes it.

The X-plates are used to move the beam *across* the screen. Usually this is done with a **time base** circuit. The time base changes the voltage across the plates so that the beam moves from left to right across the screen, and then flicks back to the start. This happens over and over again.

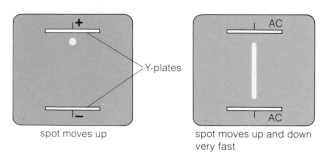

spot moves up   spot moves up and down very fast

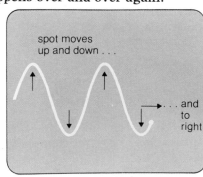

Measuring an AC voltage

With an AC voltage, the actual value of the voltage doesn't stay the same from one moment to the next.

It's the voltage which gives the 'peaks' and 'troughs' on the screen. It is sometimes useful to know the **peak voltage**. This is the maximum voltage in either the forward or backwards direction.

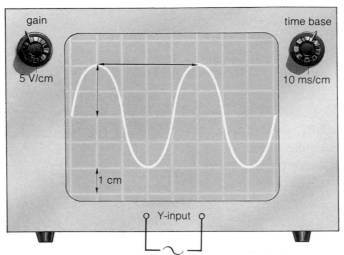

Measuring a time

An oscilloscope can be used to measure time. For example you could use it to measure the time between two peaks of an AC voltage.

To find the time between peaks:

Note the setting on the time base control.
On this oscilloscope, it is 10 milliseconds per cm.
This means that it takes 10 milliseconds for the spot to travel 1 cm across the screen.

To find the peak voltage:
Note the setting on the **gain control**. On this oscilloscope, it is 5 volts per cm. This means that the spot moves 1 cm up (or down) for every 5 volts across the Y-input terminals.

Measure the **amplitude** of the wave on the screen – that's the distance from the centre line to the peak. On this screen, it is 2 cm.

Calculate the peak voltage:
the spot moves 2 cm up or down,
there are 5 volts to every cm,
So the peak voltage is 2 × 5 volts, or 10 volts.

Measure the distance across the screen from one peak to the next. On this screen, it is 4 cm.

Calculate the time:
the spot moves 4 cm from one peak to the next,
each cm takes 10 milliseconds,
so the time between peaks is 4 × 10 milliseconds, or 40 milliseconds.

1 What is *thermionic emission?* Where is it used in a CRO?
2 Each diagram shows the front view of the X- and Y-plates of an oscilloscope. It also shows the spot on the screen.

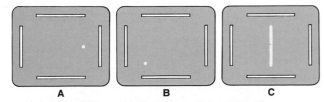

Copy diagrams A and B. Show which plates are being used to move the spot by marking them + or −.
How is the line in diagram C produced?

3 This is the screen of a small oscilloscope, *shown actual size.* An AC voltage has been put across the Y-input terminals of the oscilloscope. The control settings are also shown.

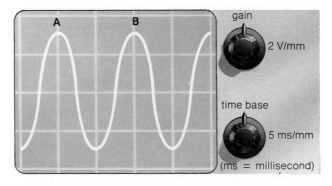

Use a millimetre ruler to measure the amplitude of the wave and the distance between peaks. Then calculate:
a the peak voltage;
b the time between peaks A and B.

5.8 Resistors and capacitors

Circuits like those in radios and cassette recorders are called **electronic** circuits. The parts fitted in these circuits are called **components**. Resistors and capacitors are components used in nearly all electronic circuits.

Resistors

Resistors keep currents and voltages at the sizes that other components need to work properly. Each resistor has a resistance measured in **ohms** (Ω). The value is marked on the side using either a colour code or a code such as '4K7' (standing for 4700 Ω).

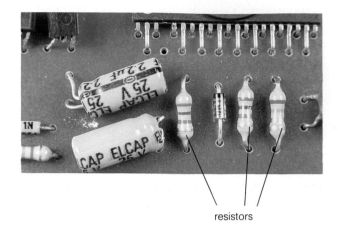

resistors

Variable resistors

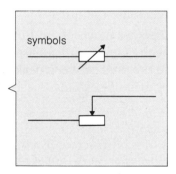

symbols

You can alter the resistance of a variable resistor by turning a spindle or moving a slider.

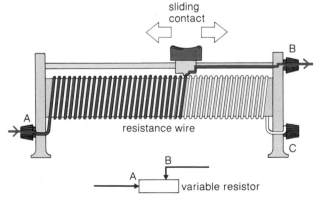

sliding contact

resistance wire

variable resistor

A **rheostat** is a type of variable resistor used to control the current flowing in a circuit.

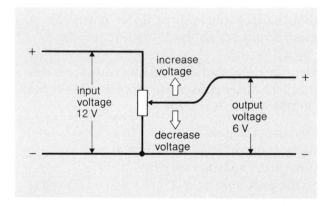

increase voltage

input voltage 12 V

decrease voltage

output voltage 6 V

This variable resistor has three terminals. It is being used as a **voltage divider** or **potentiometer**. It is passing on just part of the 12 volts put across it. Its output voltage can be anything from 0 V to 12 V, depending on the position of the sliding contact.

Capacitors

Capacitors are stores for small amounts of charge. In one way, they are like rechargeable batteries. They can take in charge, store it, and give it back later. But they store less charge than most batteries. And they only store it for a short time.

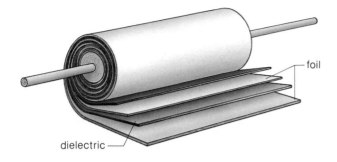

foil

dielectric

Most capacitors contain two foil sheets, rolled up like a 'swiss roll'. Sandwiched between them is a layer of insulator called a **dielectric**. Charge can't flow across the dielectric. But it builds up on the foil sheets when a voltage is put across them.

The higher the capacitance, the more charge a capacitor stores when connected across a battery.

Capacitance is usually measured in **microfarads** (**μF**).

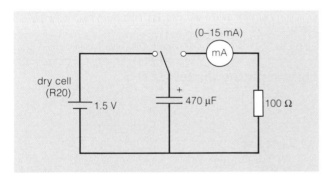

You can use this circuit to show that a capacitor stores charge. When the switch is put to the left, the battery charges up the capacitor. When the switch is put to the right, the capacitor discharges through the resistor. The flow of charge only lasts for a fraction of a second. The discharge takes longer if you use:

● a higher resistance;

● a capacitor with a higher **capacitance** value marked on the side.

Electrolytic capacitors have the largest capacitances. But their dielectrics are damaged if they are connected the wrong way round. One terminal is marked +, so that you know which way to make the connection.

symbols:

capacitor   electrolytic capacitor

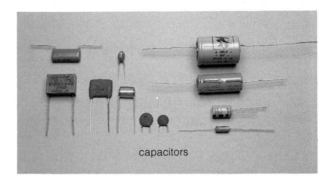

capacitors

1 Give *two* ways in which you might use a variable resistor.

2

| electrolytic capacitor | potentiometer | resistor |

Which of these:
a is used to give variable voltage output;
b stores small amounts of charge;
c can be used as a variable resistor;
d has three terminals;
e has + and − terminals which must be connected the right way round?

3

Which of these voltage dividers has an output of
a 4 V **b** 6 V **c** 0 V?
What is the output of the other voltage divider?

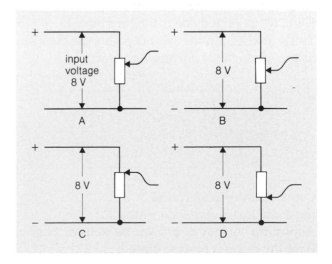

5.9 Diodes

These are **diodes**. They're found in radios, computers, and almost every other piece of electronic equipment. They have many uses, but one of the most important is in changing AC to DC (see next section).

Mostly, they are made from specially treated crystals or 'chips' of the semiconductor silicon. They are the simplest of all silicon chips.

The job of a diode is to allow current through itself in one direction, but not the other.

These circuits show what happens:

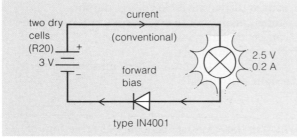

Connected this way round,
the diode has a very *low* resistance.
So a current flows, and the bulb lights up.
Here, we say the diode is **forward biased**.
The arrowhead in the symbol points the *same way* as the *conventional* current direction.

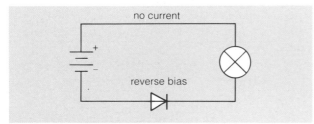

Connected this way round,
the diode has a very *high* resistance.
Almost no current flows. The bulb doesn't light.
Here the diode is **reverse biased**.

Diodes solve a problem

The problem How to wire up your circle lights so that a battery takes over when you aren't pedalling fast enough for the dynamo to light the bulbs.

The answer Use two diodes like this. (To keep things simple, only one bulb is shown):

Current can come from *either* the dynamo *or* the battery, depending on which has the higher voltage. But the dynamo can't push current through the battery. And the battery can't push current through the dynamo. The two diodes stop that happening.

Light-emitting diodes (LEDs)

These diodes give off light when they are forward biased. They are often used as indicator lights on videos and cassette players.

LEDs can't handle large currents. To protect them from damage, a high resistance is usually connected in series.

LEDs can be used to display numbers. Seven LEDs, lighting up in different combinations, can give you any number from 0 to 9 (see also page 226).

Testing a diode

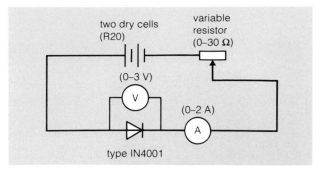

You could use this circuit to find out how the current through a diode changes with the voltage across it. The voltage is set at different values by adjusting the variable resistor, and the current measured each time. Here, the diode is forward biased. For reverse bias, you just change the battery connections round.

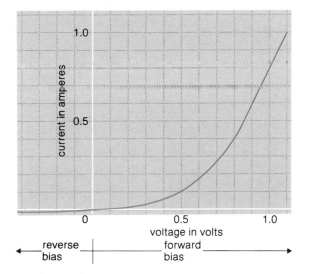

These are the results you might get, in the form of a graph. To find the resistance of the diode at any point, use the equation:

resistance = voltage ÷ current
(ohms)   (volts)   (amperes)

When reverse biased, this diode has a very high resistance.
When forward biased, its resistance becomes less as the current rises.

Keep within limits Too much current, or too much reverse voltage, and a diode is damaged. The diode in this experiment isn't designed for currents of more than 1 A. Its reverse bias limit is about 50 V.

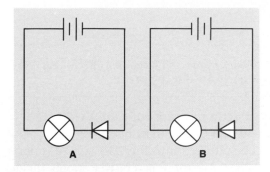

In which of these circuits:
a does the bulb light up;
b does the diode have a high resistance;
c is the diode forward biased?

2 The radio is powered by a battery. But if the battery is connected the wrong way round, parts inside the radio may be damaged.

Design a circuit so that current can flow if the battery is connected as shown, but is blocked if the battery is connected wrongly.
3 Redraw the battery and dynamo circuit on the opposite page so that two bulbs are supplied with current instead of one. Assume that you are using a 6 V battery and two 6 V bulbs.
Explain why, when the dynamo isn't generating, the battery pushes all its current through the bulbs and none through the dynamo.
4 Using readings from the graph on this page, copy and complete the following table:

Diode forward biased		
Voltage across diode	Current in A	Resistance in ohms
0.5 V	?	?
1.0 V	?	?

5.10 Power supplies

AC – 'backwards and forwards' alternating current.
DC – 'one way' direct current.

This cassette recorder runs on 9 volts DC.
But you don't have to fit batteries.
It can be plugged into the 230 volt AC mains.
This is because there is a **power supply** inside.
It reduces the voltage, and changes the AC to DC.

Changing AC to DC

Changing AC to DC is called **rectification**.
It is done with diodes.
Diodes used in this way are called **rectifiers**.

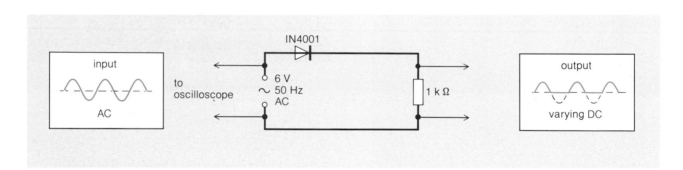

This circuit changes AC to DC. The diode lets the forward parts of the alternating current through, but stops the backward parts. This means that current only flows one way through the resistor.

An oscilloscope can be used to show what the diode does to AC. When the oscilloscope is connected across the resistor, the bottom half of the AC waveform is missing. The current is flowing in surges, with short periods in between when there isn't any current. This is **half-wave rectification**.

Smoothing with a capacitor

The 'half-wave' current from a diode is far too jerky for most electronic equipment. But the flow can be smoothed out with a capacitor. This collects charge during the surges, and releases it when the current from the diode stops. This gives a much smoother flow of current through the resistor.

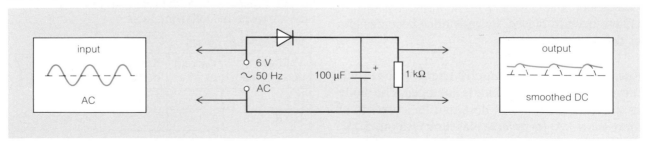

A power supply

This is a simple power supply. It contains a transformer, four diodes, and a capacitor. It changes the mains AC into low-voltage, smoothed DC:

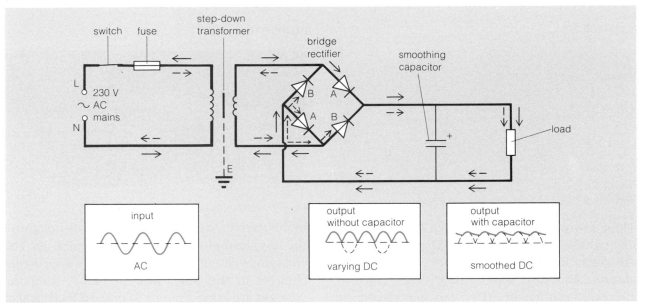

The transformer steps down the mains voltage. In some power supplies, for safety, a switch and a fuse are fitted in the live wire, and the core of the transformer is earthed.

The four diodes together make a **bridge rectifier**. This changes the AC to DC. But instead of *blocking* the backwards parts of the alternating current, it actually *reverses* them so that the whole flow is forwards. The result is **full-wave rectification**.

The bridge rectifier works like this: When the AC is flowing forwards (→), diodes A conduct, and diodes B block. When the AC is flowing backwards (←−), diodes B conduct, and diodes A block. Follow the arrows and you will see that the current from the bridge rectifier always flows the same way.

The capacitor smooths the current flow.

The load is a simple resistor in the circuit above. But it could be a radio, or a cassette recorder, or any other piece of equipment needing a low-voltage DC supply.

1 On an oscilloscope, which of these waveforms is produced by:
a AC;
b half-wave rectification of AC;
c full-wave rectifiction of AC;
d rectification of AC, with smoothing?
2 Read pages 166 and 167. Then explain why, in a power supply:
a the switch and the fuse should be placed in the live wire;
b the core of the transformer should be earthed.

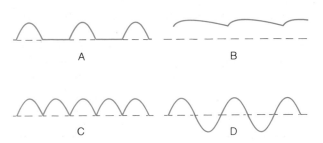

5.11 Transistors

Transistors...

These are **transistors**. Inside each case is a specially treated crystal of silicon, with three connecting leads attached. Transistors are used in TVs, CD players, and just about every other piece of electronic equipment. They can join circuits so that the current through one controls the current through the next.

amplifying...

In this stereo system, transistors **amplify** (magnify) the tiny, changing voltage from the pick-up. The changing voltage in one circuit causes even bigger changes in the next. And so on ... until the changes are large enough to make a loudspeaker cone vibrate.

A switched-off transistor

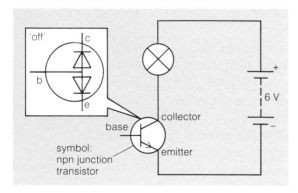

The transistor in this circuit is an **npn junction transistor**. A circuit symbol has been used to show it. It has three terminals – called the **emitter**, the **base**, and the **collector**.

Connected like this, the transistor behaves like two diodes back-to-back. The bottom diode can conduct, but the top diode has a blocking effect. No current can flow. The bulb doesn't light. The transistor is 'switched off'.

Switching the transistor on...

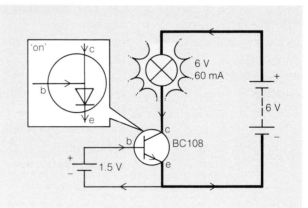

Put a small voltage across the base and emitter like this, and a tiny current flows through the base. This current completely alters the way the transistor behaves. The blocking effect vanishes, the transistor starts to conduct, and the bulb lights up. It takes a base current of less than 1 mA to 'switch on' the transistor so that the bulb lights.

The transistor won't switch on unless the + and − connections are the right way round. In each circuit, the conventional current direction must be the same as the arrow direction in the symbol.

switching...

In automatic washing machines, transistors are used as switches. They control the circuits which operate the wash program.

Transistor switches do a similar job to the relay on page 181. You can find out more about them in the next section.

and shrinking

These are **integrated circuits** or **ICs**. They contain many complete circuits, with transistors, resistors, connections, and other components all formed on a tiny chip of silicon only a few millimetres square. Most of the transistors in a washing machine are in 'microchips' like these.

and another way of switching it on

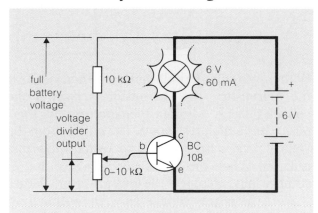

You don't need a second battery to switch a transistor on. With a voltage divider, you can use a proportion of the main battery voltage instead.

Here, a resistor and a variable resistor are being used as a voltage divider. If the sliding contact is at the bottom, the output from the voltage divider is zero. This keeps the transistor switched off. If the sliding contact is moved upwards, the output from the voltage divider rises. When it passes 0.6 V, the transistor switches on and the bulb lights.

1 In the circuit, the bulb is glowing brightly.

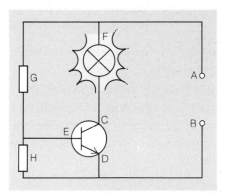

a Which letter, A to E, stands for each of these?
base   collector   battery+   emitter?
b Draw the circuit diagram. Mark on arrows to show the conventional current direction in each part of the circuit.
c What do the components G and H do?
d What would happen if you replaced G with a short piece of connecting wire? Explain why.

5.12 Transistor switches

The doors open when a customer approaches.
The entrance lights come on when darkness falls.
When the manager sets the overnight burglar alarm, she has time to leave the building before it switches on.
If there is a fire, an alarm bell sounds.
These all happen automatically.
They're all controlled by transistor switches.
The circuits below show the basic principles.

A light-operated switch

This circuit uses a **light-dependent resistor (LDR)**. This is a special type of resistor whose resistance falls when light shines on it.

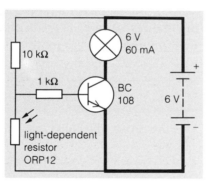

With this circuit, the bulb lights up when the LDR is put in the dark.

The LDR is part of the voltage divider.
In daylight, the LDR has a low resistance, and a low share of the battery voltage. The output from the voltage divider isn't enough to switch the transistor on.

In darkness, the resistance of the LDR rises. This gives the LDR a much larger share of the battery voltage. Now, the output from the voltage divider is more than the 0.6 V needed to switch the transistor on. So the bulb lights up.

If you replace the upper resistor with a variable resistor, you can alter the 'darkness level' needed to switch the transistor on.

If you swap round the LDR and the upper resistor, the bulb will then be 'off' in the dark and 'on' in the light.

A heat-operated switch

This circuit uses a **thermistor**. This is a special type of resistor whose resistance falls when it is heated.

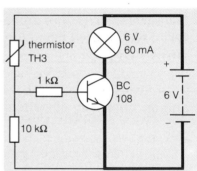

With this circuit, the bulb lights up when you heat the thermistor. The thermistor is part of the voltage divider. At room temperature, the thermistor has a high resistance. The output from the voltage divider isn't enough to switch the transistor on. When the thermistor is heated, its resistance falls. This gives the lower resistor a larger share of the battery voltage. Now, the output from the voltage divider is more than the 0.6 V needed to switch the transistor on. So the bulb lights up.

If you replace the lower resistor with a variable resistor, you can alter the temperature needed to make the transistor switch on.

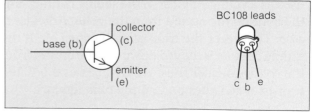

A time-operated switch

This circuit uses a **capacitor** to delay the switching on of the transistor. When the battery switch is closed, you have to wait several seconds before the bell rings.

To reset the circuit, you open the battery switch. Then you discharge the capacitor by closing the switch across it.

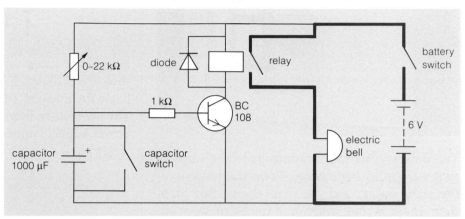

The capacitor is part of the voltage divider. When you close the battery switch, the capacitor slowly charges, and the voltage across it rises. It takes several seconds for the voltage to reach the 0.6 V needed to switch the transistor on.

It takes even longer if you use:

- A higher resistance setting on the variable resistor. This makes the capacitor charge more slowly.
- A capacitor with a higher capacitance value.

This circuit has an important extra feature. The transistor doesn't switch on the bell directly. Instead it switches on a relay (see page 181). The relay then switches on the bell. The advantage of this is that the current in the bell circuit doesn't have to flow through the transistor. So a higher current can be used.

The diode protects the transistor from currents generated when the relay contacts open.

1 This chart gives information about four transistor switch circuits, and what is needed for the transistor to be switched ON or OFF:

	Transistor ON	Transistor OFF
Circuit A	darkness	bright light
Circuit B	bright light	darkness
Circuit C	high temperature	low temperature
Circuit D	low temperature	high temperature

which circuit is most like the one that would be used to:
a open an automatic door when a light beam is cut;
b set off an alarm when a fire starts;
c switch on a car parking light at dusk;
d switch on a heater when the temperature in a greenhouse falls?

2 Design a transistor switch circuit with the following features:
- an electric motor is switched on automatically in bright light;
- the motor is switched on by a relay.
What is the advantage of using a relay in this type of circuit?
3 In the circuit at the top of the page, what would be the effect of:
a setting the variable resistor to a lower resistance value?
b replacing the 'normally open' relay with a 'normally closed' one.
Can you think of a practical use for the circuit in b?

5.13 Logic gates

The door of this safe is controlled by groups of switches called **logic gates**. Press the buttons in the right order, and the door opens. But press them in the wrong order, and the alarm goes off. The electronic gates needed to do this job are quite complicated.

The gate used in the circuit below is much simpler – just two switches in a box:

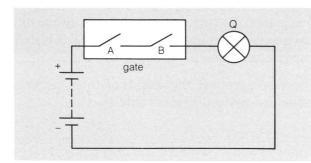

With this gate:
If both switches are closed the bulb comes ON.
But if either switch is open, the bulb stays OFF.

You can use a **truth table** to show the result of every possible switch setting. The table uses two **logic numbers**, 0 and 1:
Logic 0 stands for OFF. Logic 1 stands for ON.

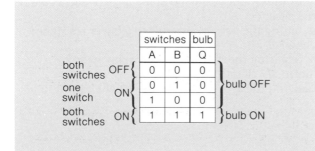

		switches		bulb	
		A	B	Q	
both switches	OFF {	0	0	0	
one switch	ON {	0	1	0	} bulb OFF
		1	0	0	
both switches	ON {	1	1	1	} bulb ON

Logic gates can be grouped together on a single chip. Below, are three types of logic gate, along with their truth tables. In the tables:
Logic 0 stands for OFF. Logic 1 stands for ON.
The gates have been drawn using symbols.
Power supply connections aren't shown.

AND gate

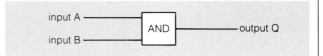

This AND gate has one output and two inputs.
The output is only ON if *both* inputs are ON.
In other words, Q is ON if both A *AND* B are ON.
This is the truth table for the AND gates:

		inputs		output	
		A	B	Q	
both inputs	OFF {	0	0	0	
one input	ON {	0	1	0	} output OFF
		1	0	0	
both inputs	ON {	1	1	1	} output ON

OR gate

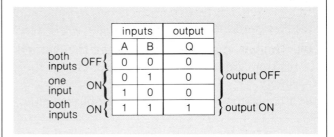

This OR gate has one output and two inputs.
The output is ON if *either* input is ON.
In other words, Q is ON if either A *OR* B is ON.
This is the truth table for the OR gate:

		inputs		output	
		A	B	Q	
both inputs	OFF {	0	0	0	} output OFF
one input	ON {	0	1	1	
		1	0	1	} output ON
both inputs	ON {	1	1	1	

NOT gate

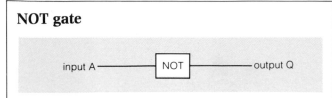

A NOT gate has one output and one input.
The output is ON if the input is OFF, and vice versa.
This type of gate is called an **inverter** gate.
This is the truth table for a NOT gate:

	input	output	
	A	Q	
input OFF {	0	1	} output ON
input ON {	1	0	} output OFF

Safe with logic

The manager of a camera shop wants an electric lock on her storeroom door. She needs a system without combinations or numbers to remember, so that any member of staff can unlock the door quickly. She decides on a two-switch system. If a hidden switch is turned on first, a main switch will open the door lock. But if the hidden switch is left off, the main switch will turn on an alarm instead.

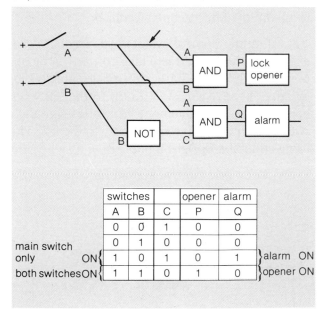

switches			opener	alarm	
A	B	C	P	Q	
0	0	1	0	0	
0	1	0	0	0	
1	0	1	0	1	} alarm ON
1	1	0	1	0	} opener ON

main switch only ON (rows with 1 0 1 0 1); both switches ON (1 1 0 1 0)

This is the system of logic gates she decides to use, together with its truth table. To keep things simple, complete circuits aren't shown. Nor are the connections to the power supply.

With most gates, an input is made ON by applying a small positive voltage (such as +5 V). That is why scientists often talk about inputs being HIGH or LOW (voltage) rather than ON or OFF.

1

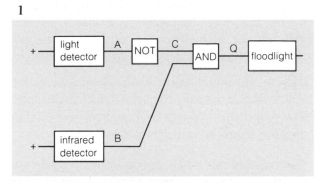

This is the security lamp system on the front of a house. Under certain conditions, a floodlight is switched on automatically.
Daylight turns the light detector ON. Body heat turns the infrared detector ON when anyone approaches.
a Write down the truth table for the system using column headings A, B, C, and Q.
b Use the table to work out what happens if someone approaches when it is dark.
c Use the table to work out what happens if someone approaches when it is daylight.

2 If you combine an AND gate with a NOT gate, the result is a **NAND gate**.
If you combine an OR gate with a NOT gate, the result is a **NOR gate**.

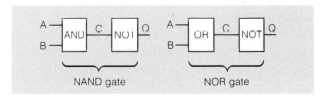

For each gate:
Write down the truth table, using column headings A, B, C, and P.
Use the table to work out whether the output is ON or OFF when:
a both inputs are ON; **b** both inputs are OFF;
c ony one input is ON.
3 When the storeroom lock on the left was installed, a mistake was made. The NOT gate was fitted in the top input line (*see arrow*) instead of the bottom one.
a What happened when the manager turned the hidden switch ON?
b What happened when she then turned the main switch ON?
(Drawing a truth table may help you to work out the answers.)

223

5.14 Sensors and systems

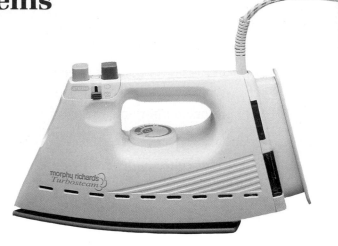

Sensors

You turn ordinary switches on and off with your fingers. However, logic gates and other transistor switches are turned on and off by tiny electric currents. They are electronic switches.

The current which controls an electronic switch comes from a **sensor**. Sensors detect heat, light, sound, pressure from your finger, or some other **stimulus** from the outside world. For example:

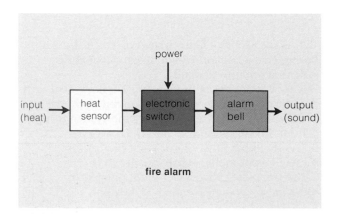

fire alarm

The flow diagram above shows a fire alarm system which uses an electronic switch. The system has an **input** (heat). The result is an **output** (sound).

Electronic switches can only handle tiny currents. So they cannot directly switch on powerful things like electric motors and alarm bells. To get round this problem, a **relay** is used (see page 181). The electronic switch turns on the relay, and this switches on the more powerful circuit.

Control systems and feedback

A switch is a simple form of **control system**. It controls the operation of a circuit.

Some control systems are automatic. For example, the temperature of an electric iron is regulated by a **thermostat** (see page 83). This contains a switch controlled by a heat sensor. It switches the iron OFF if it gets too hot, then ON when it cools down.

For the thermostat to work, some of the heat from the iron must be fed back to the heat sensor. In

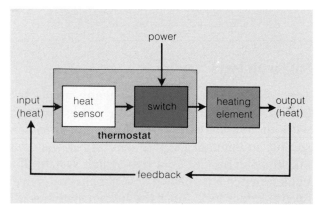

other words, some of the output must be fed back to the input. This is called **feedback**.

All automatic control systems use feedback. For example, when you ride a bike you are using one of your body's automatic control systems to keep balance. If, say, you start to tip to the left, your body senses the movement and sends signals to your arm muscles to steer slightly to the left. This brings the bike upright again.

Every automatic control system has these parts:

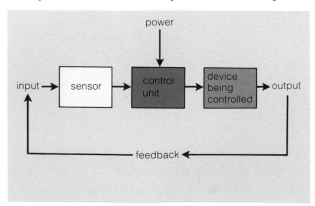

A bistable

Note HIGH = ON = logic 1, LOW = OFF = logic 0

The arrangement below is called a **bistable**. Two NOR gates have been cross-coupled so that the output of each is fed back to one input on the other. This arrangement has a special feature: with both inputs LOW, there are *two* possible output states ('bi' stable states).

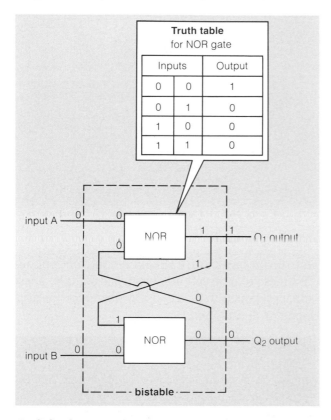

With both inputs LOW:
- *either* the top output is HIGH and the bottom output is LOW (as shown above);
- *or* the top output is LOW and the bottom output is HIGH.

The bistable can be made to flip from one state to the other, and so on. For example:

If, in the diagram above, the top input is made HIGH, the bistable flips to the other output state. The input signal does not have to be maintained. One brief positive pulse is all that is needed to make the change. The bistable 'remembers' its last output state. This feature makes bistables very useful in computers, where data is stored in memory as a series of HIGH and LOW states.

A latch

A bistable with just one of its outputs in use is called a **latch**. By applying positive pulses to its inputs, the output can be made to go ON and OFF alternately.

Below, a latch is being used to control an electric motor. Normally, each input is LOW (0 V), but it goes HIGH (+5 V) if a button switch is pressed. One switch starts the motor, the other stops it. The touch buttons used to start and stop a video recorder work in this way.

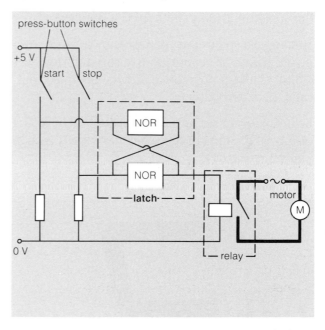

1 Look at the two transistor circuits on page 220. For each one, say what is the *input*, what is the *output*, and what is being used as a *sensor*.
2 To control the temperature inside a refrigerator, the motor which pumps the refrigerant round has to be switched on and off automatically.
a What device switches the motor on and off?
b How is this an example of *feedback*?
c Draw a flow diagram for the control system.
3 Look at the bistable diagram on this page. Redraw it to show the bistable in its other output state. Someone claims that a bistable has a 'memory'. What do you think they mean by this?
4 Which of the following jobs do you think could be done using a *latch*? Explain your answer(s).
a Switching a calculator on and off with separate buttons.
b Keeping a heater at a steady temperature.
c Switching on a light when a team member presses their button in a TV quiz show.

5.15 Dealing with signals

Analogue and digital

The voltmeters on the right have different types of display:

One meter has a pointer which moves up a scale. It has an **analogue** display. It is using some feature (in this case, the angle of the needle) to represent the voltage.

The other meter shows the voltage as a number. It has a **digital** display (a 'digit' is a number).

Analogue displays are good for showing measurements which are continuously varying. However digital readings are more easily handled electronically, by a computer.

Below, you can see how a digital display shows the number '3'. Some of the segments are ON (black) and some are OFF.

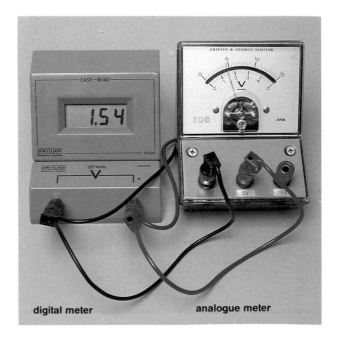

digital meter   analogue meter

Sending sound

When you speak into a telephone, the microphone changes your speech into electrical signals. These can be transmitted over long distances by analogue or digital methods.

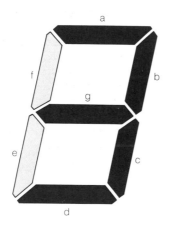

To display '3', the segments must be as follows:

a	b	c	d	e	f	g
ON	ON	ON	ON	OFF	OFF	ON

Using logic numbers, this can be written:

1	1	1	1	0	0	1

For this, the gate voltages are:

HIGH HIGH HIGH HIGH LOW LOW HIGH

This sequence can be stored using bistables.

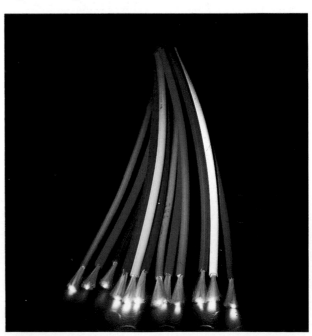

Optical fibres can carry digital signals in the form of light pulses

The electrical signals from a microphone are a tiny, changing current which varies just like the incoming sound vibrations. The graph on the right shows how the current might vary over one brief fraction of a second. It is an analogue signal.

The problem with analogue signals is that they are affected by electrical disturbances on the line, called **noise**. The effect of noise is greatly reduced if the signal is changed into a digital signal. To do this, the analogue signal is **sampled** electronically thousands of times every second. This means that the level of the signal (the height on the graph) is measured at regular, short intervals. Then the measurements are changed into **binary codes** (numbers which use only 1s and 0s). These are transmitted as a series of pulses (HIGHs and LOWs) as on the right.

Sampling level	Binary code	Pulses (voltage sequence)		
0	000	LOW	LOW	LOW
1	001	LOW	LOW	HIGH
2	010	LOW	HIGH	LOW
3	011	LOW	HIGH	HIGH
4	100	HIGH	LOW	LOW
5	101	HIGH	LOW	HIGH
6	110	HIGH	HIGH	LOW
7	111	HIGH	HIGH	HIGH

On the graph, only 7 sampling levels have been used. In a real system, there are 256 levels, and the sampling rate is much faster.

Modern telephone systems use **optical fibres** for transmitting digital signals over long distances (for more on optical fibres, see page 117). A tiny laser puts pulses of light into one end of the fibre. At the other end, a sensor changes the pulses into electrical signals. Optical fibres can carry more calls than electric wires, and the signal does not weaken so much with distance.

At the receiving end, the digital signal is changed back into an analogue signal before being sent to the telephone earpiece.

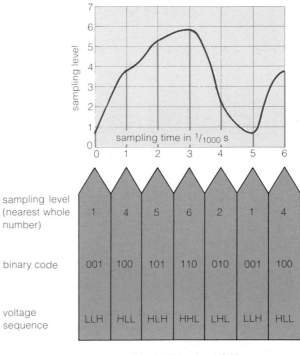

(H = HIGH   L – LOW)

1 A wrist-watch can have either an *analogue* or a *digital* display.
What is the difference between the two types?
What are advantages of each type?
2 Telephone systems often use digital signals for sending speech. Give *two* advantages of using digital signals instead of analogue ones.
3 The graph below shows an analogue waveform.
a Write down the sampling level at each 1/1000 s.
b Change the levels into binary codes.
c How can this binary code sequence be transmitted in the form of electrical signals?

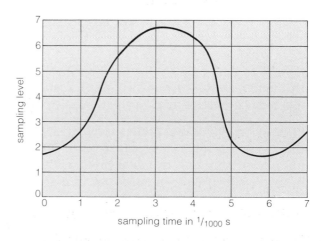

5.16 Machines that think

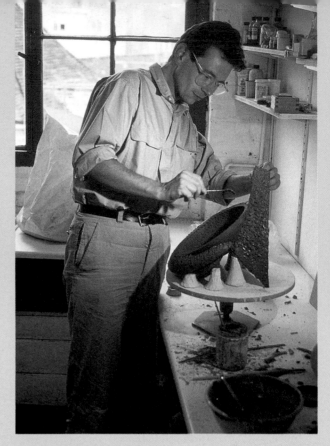

rules and learn from their mistakes. But they have a long way to go. Microchips still can't match humans for ideas.

New hope for the disabled

Electronics may help paralysed people to walk again, with the help of a strap-on walking frame. The latest frames are driven by compressed air, and a microchip controls the walking action.

Human operators are skilled and creative but machines are faster and they never get tired.

In industry, more and more machines are being run by microchips. They can control anything from paint spraying to metal drilling. But they have to follow a set of rules put into them by people. They can't work without a **program**.

Now, scientists are designing microchips with 'artificial intelligence' – chips that set their own

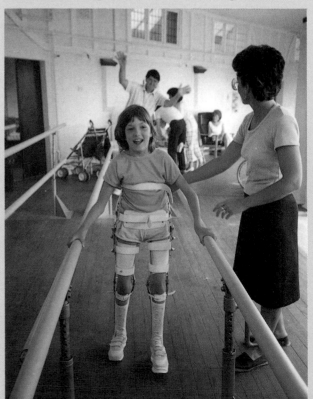

Reducing the risk

Air crashes don't happen very often. But when they do, it is usually because of human error. In an emergency, the crew has to make many decisions very fast. One wrong action can bring disaster.

Now, microchips are making flying safer. In the latest systems, microchips analyse the crew's actions, give a warning if they aren't safe, and take over control if the warning is ignored.

Could microchips declare war?

At least one American scientist thinks so. He claims that computers are now being left to decide when nuclear missiles should be launched. He thinks that this is against the law. And he is taking USA defence chiefs to court to prove it.

The defence chiefs say that if the USA is attacked, computers could launch a counter-attack more quickly than human beings. But they claim that this wouldn't happen. The President would still make the final decision.

Microchips down on the farm

Electronic tractors will work best in large fields with no obstacles like trees and hedges to get in the way. They won't be welcomed by people who like the countryside the way it is. And they won't be welcomed by Britain's 70 000 unemployed farm workers.

Soon, this tractor won't need a driver. Microchips will control it, along with the ploughing, sowing spraying, and harvesting. This should help to put food production up and keep costs down.

Electronic farms are likely to mean greater food production at lower cost. But how many disadvantages can you think of?

In airliners, microchips will take over more and more of the jobs done by the crew. But passengers are still likely to want a crew on board. Can you think of the reasons why?

A microchip can control the heating system in your house. What other household tasks do you think could be done by microchips?

Questions on Section 5

1 There are two types of electric charge, positive (+) and negative (−). Both types are found in the atom.

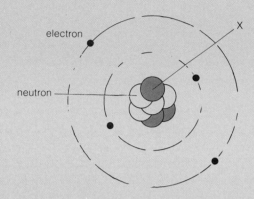

electron

neutron

X

a What type of charge does an electron have?
b What is particle X called?
c What type of charge does particle X have?
d How many particles of type X would you expect the atom above to have? Explain your answer. (Note: the diagram does not show all the particles of type X.)
e When a current flows through a wire, it is some of the particles from the atoms which are moving along. Which particles?

2 In a papermaking plant, a radioactive source and detector can be used for testing the thickness of the paper as it is rolled out.

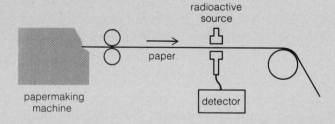

radioactive source

paper

papermaking machine

detector

The radioactive source in the diagram above gives out beta particles.
a What are beta particles?
b What device could be used as a detector?
c If the paper is thicker than normal, what effect will this have on the reading on the detector?
d Why should the radioactive source be a beta source, rather than one that gives out alpha particles or gamma rays?

e Some people could be worried that the radiation might 'make the paper radioactive'. What answer would you give them?

3 The diagram below shows the basic features of one type of nuclear power-station.

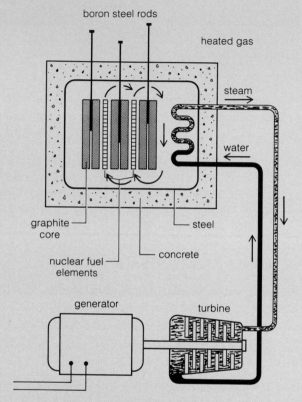

boron steel rods

heated gas

steam

water

graphite core

steel

nuclear fuel elements

concrete

generator

turbine

a What name is given to the process which releases heat in the core of the reactor?
b How is the heat output controlled?
c How is the heat used to make electricity?
d Why is the core surrounded by a steel pressure vessel and thick concrete?
e Compared with a conventional fuel-burning power-station:
  i what are the advantages of a nuclear power-station?
  ii what are the disadvantages?
f Strontium-90 is a radioactive waste product. It has a half-life of 28 years. When the radiation from a small sample of strontium-90 is measured, the count-rate is found to be 800 counts per second. What would you expect the count-rate to be
  i after 28 years?
  ii after 56 years?
  iii after 112 years?

g The amount of radiation from a small sample of strontium-90 may be very low, yet dust containing that strontium might still be extremely harmful to humans and other living things. Explain why.

4 The diagram below shows a cathode ray oscilloscope (CRO). Electrons leaving the cathode are accelerated towards the anode. They pass between two sets of deflector plates before striking the screen.

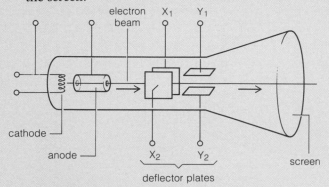

a What makes the cathode give off electrons? What is the effect called?
b What makes the electrons accelerate from the cathode towards the anode?
c How can the beam be made to bend upwards so that it strikes the screen nearer the top?

5 The devices below are all used in electronic circuits.

diode   capacitor   LED

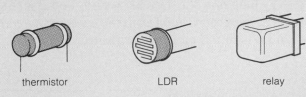

thermistor   LDR   relay

Which of the above is best described by each of the following statements?
a Glows when a tiny current flows through it.
b Links two circuits so that a small current in one can switch on or off a larger current in the other.
c Has a resistance which falls when it is heated.
d Has a resistance which falls when light shines on it.
e Allows current to pass through in one direction only.

6 The device below is called a voltage divider (also known as a potential divider). It is used in many electronic circuits.

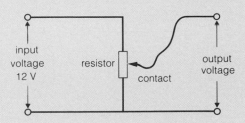

a Draw diagrams to show where the sliding contact should be positioned to give an output voltage of:
  i 12 V   **ii** 6 V   **iii** 3 V
b Give one example of the practical use of a voltage divider.

7 The box in the diagram below is a simple form of logic gate.

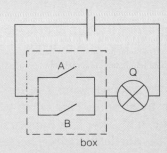

Truth table		
Inputs		Output
A	B	Q
O	O	
O	1	
1	O	
1	1	

a Copy and complete the truth table for the gate.
b Decide whether the gate is an *AND* gate, an *OR* gate, or a *NOT* gate.

8 The meters below are both displaying a voltage reading:

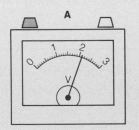 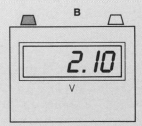

a Which one is giving the data in digital form?
b What is the other form of data called?
c In which form could the data most easily be transmitted along an optical fibre? Explain how this could be done.

6.1 Sun and seasons

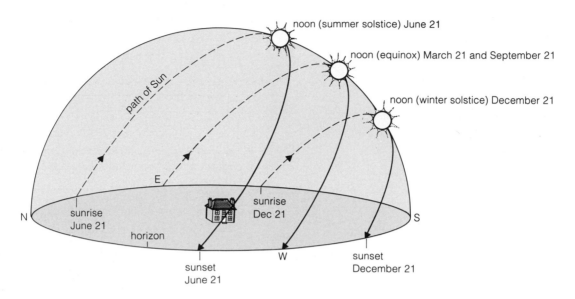

The diagram above shows how the Sun appears to move across the sky on four key days of the year, from one place on the surface of the Earth.

The Earth's surface gains most heat when the Sun is above the horizon for most hours, *and* when it reaches the highest angle in the sky. This happens around June 21st in the northern hemisphere, and December 21st in the southern hemisphere. This **Midsummer Day** is sometimes called **summer solstice**. Solstice means 'Sun standing still'.

The surface gains the least heat exactly half a year after summer solstice, when the Sun is at its lowest angle above the horizon *and* stays above the horizon for the fewest hours. This is **Midwinter Day**, or **winter solstice**.

Between the two solstices are the two **equinox** These are the days on which daytime and nigh time are of equal length.

Solar heating

The Sun is a star. Other stars look much smaller because they are much further away. The Sun radiates huge amounts of energy, a tiny fraction of which falls on the Earth.

The diagram below shows why some parts of the Earth's surface are hotter than other parts. The energy reaching the area between A and B is the same as between C and D, but the energy between C and D is much more concentrated.

Stonehenge, in Wiltshire, was built about 5000 years ago. In 1771, Dr John Smith studied the positions of stones on the outer circular bank. He put forward the hypothesis (idea) that two of the stones formed a line which pointed to the sunrise position at summer solstice. In 1846, Revd Edward Duke suggested that two more of the stones on the opposite side of the circle pointed to the sunset position at winter solstice.

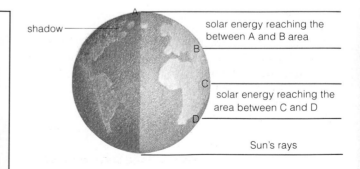

Day, night, and seasons

The Earth moves in an almost circular orbit round the Sun, taking just over 365 days to complete one orbit. The Earth also spins on its axis at the rate of one turn every 24 hours. The axis is tilted, and it is this which causes the different seasons.

It is easier to see why the seasons exist by making the model below. Hold the rod so that it is about 20° from vertical. (The exact figure is 23.5°. You could model this by mounting the rod in a wooden block measured to this angle.) Mark a line around the 'Equator', and a dot where you live.

Set your model with its tilt towards a model 'Sun' – a small electric bulb about 0.5 m away. Use a 0.5 m string to keep the distance fixed as you slowly move your 'Earth' around the 'Sun'. Move around anticlockwise, making sure that you keep the rod tilting in the same direction all the time.

As you rotate the model Earth, your dot keeps moving from light (daytime) to shadow (night-time) and back into the light again. Also, because of the tilt, your dot gets different lengths of daytime and night-time as the Earth moves round the Sun.

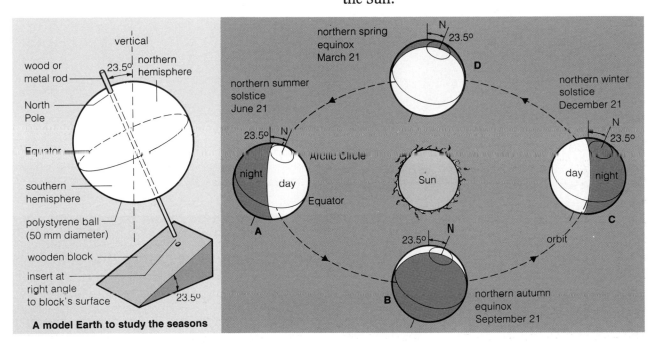

A model Earth to study the seasons

1 What do we call the time it takes the Earth:
a to make one orbit of the Sun?
b to rotate once?
2 In the diagram on the left, where is the Sun's energy least concentrated?
3 Why is it colder at the North and South Poles than at the Equator?
4 What date is the winter solstice in:
a the northern hemisphere?
b the southern hemisphere?
5 On the days of the two equinoxes, daytime and night-time are equal in length. For how long does the Sun appear in the sky on those days?
6 On which day of the year will the Sun give your part of the Earth the *least* energy?

7 On which day of the year will the Sun give your part of the Earth the *most* energy?
8 Look at the diagram above. At which positions in the Earth's orbit, do all regions get daytime and night-time of equal lengths?
9 Decide whether each of these statements is TRUE or FALSE:
a When the Earth is at position C, it is winter in the southern hemisphere.
b When the Earth is at position B, it is spring in the southern hemisphere.
c When the Earth is at position A, there will be 24 hours of daylight at the North Pole.
10 Why do you think the seasons were even more important to people 5000 years ago than today?

6.2 The Moon in orbit

The Moon moves in a circular orbit around the Earth, held there by the gravitational pull between it and the Earth. We see the Moon because its surface reflects sunlight. Depending on the position of the Sun, different amounts of the sunlit part are visible. These different views of the Moon are called **phases**. The Moon takes 27 days to orbit the Earth, but a complete sequence of phases takes about 28 days. That is because the Earth also changes position as it orbits the Sun.

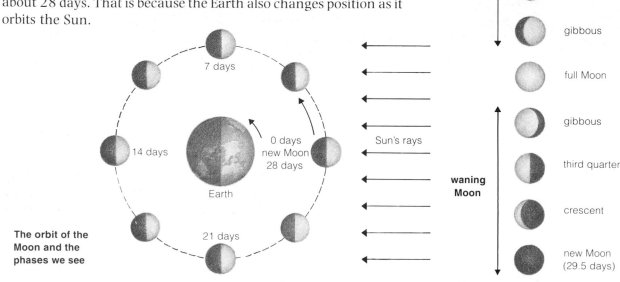

The orbit of the Moon and the phases we see

The Moon spins at the same rate as it orbits — once every 27 days — so it always keeps the same face towards the Earth.

The craters on the Moon are mainly caused by the impact of large meteorites over a billion years ago. The 'seas' are not seas at all, but fairly flat areas of basalt rock.

Diameter 3480 km
Mass 7.4×10^{22} kg
(0.012 of Earth)
Typical surface temperature
$-150\,°C$ to $120\,°C$
Atmosphere None
On the picture, coloured dots show where spacecraft landed.

Eclipses

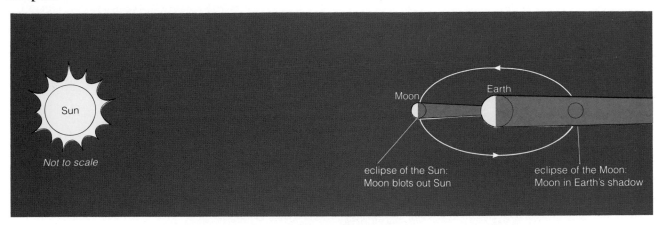

Sun

Not to scale

Moon

Earth

eclipse of the Sun:
Moon blots out Sun

eclipse of the Moon:
Moon in Earth's shadow

The Moon is much smaller and closer than the Sun, but it *looks* the same size from Earth. As the Moon orbits the Earth, it sometimes passes exactly between us and the Sun. Then, the Sun's disc is blotted out – there is an **eclipse of the Sun**.

Gravity and tides

On Earth, the Moon's gravity pulls water into two tidal bulges. As the Earth rotates, each place passes in and out of a bulge about twice a day. So the sea-level rises and falls and we have **tides**.

The high tide on the side of the Earth *facing* the Moon is easy to explain by the Moon's gravity pulling the sea. The high tide on the far side happens because the water on that side is pulled by the Moon's gravity less than the Earth itself is pulled. So the water gets 'left behind' a little.

Eclipses of the Sun do not happen very often because the Earth's orbit is tilted. More frequently, the Moon's orbit takes it into the Earth's shadow. Then the Moon darkens – there is an **eclipse of the Moon**.

1 How long does it take the Moon to orbit the Earth?
2 Why do we always see the same side of the Moon?
3 Why is the Moon sometimes seen as a crescent rather than a full disc?
4 Approximately what length of time is there between a new Moon and a full Moon?
5 Describe what happens:
a during an eclipse of the Sun;
b during an eclipse of the Moon.
6 What causes tides? Why are there two high tides per day?

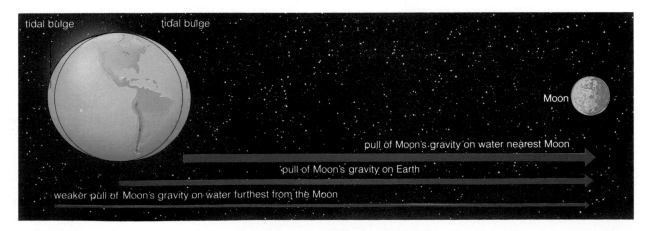

tidal bulge

tidal bulge

Moon

pull of Moon's gravity on water nearest Moon

pull of Moon's gravity on Earth

weaker pull of Moon's gravity on water furthest from the Moon

6.3 Patterns in the sky

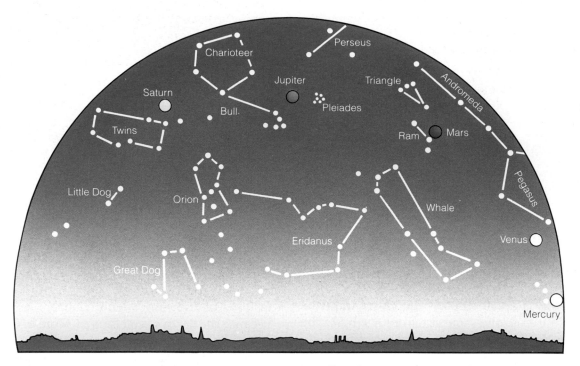

For thousands of years, people have looked at the sky at night. The stars appeared to make patterns which they interpreted as animals, people or objects. Anyone watching the sky for a few hours would see that the pattern of stars appears to move around one special star.

Imagine a black umbrella with bright spots painted on the inside. If you turn the umbrella anticlockwise around its handle, the spots move around a point in the middle. This is how the stars appear to move around the Pole Star, *Polaris*.

Really, the sky is still and the Earth is turning under it. Imagine keeping the umbrella still, but turning yourself clockwise.

The star patterns, called **constellations**, are still used by scientists to make star maps.

People noticed that five lights in the sky moved about. During one night, they seemed just like fixed stars, but after a few weeks they had moved to different positions. These moving lights became known as **planets** from a Greek word meaning 'wanderer'. Some cultures named them after various gods and goddesses.

Long ago, people thought the stars were all the same distance away, like the umbrella spots. Then the following discovery changed this opinion.

The distance between all the stars looks fixed, but by looking at them carefully, and then again six months later, scientists could see that some stars appeared to have moved a little. This is because the stars are now being seen from the other side of the Earth's orbit around the Sun. The effect is similar to the changing view you see if you look out of the side window of a moving car. Trees and buildings close to you appear to 'move' more than those which are further away.

The apparent movement is called **parallax**, and it can be used to calculate the distance to the stars from the Earth. The car's journey is like the Earth's journey from one side of its orbit to the other.

Parallax measurements show that the nearest star (apart from the Sun) is nearly 43 million million kilometres away. This can be written as 4.3×10^{13} km. The star is called *Proxima Centauri*.

Light rays travel at a speed of 300 000 km/second. The distance they travel in one year is called a **light year**. Proxima Centauri is 4.5 light years away, so its light has taken 4.5 years to reach us. For comparison, the Sun is about 8 light minutes away from Earth (144 million kilometres).

On clear nights, a faint, ragged band of light can be seen across the sky. It is made up of millions of stars, so far off that we cannot detect parallax. It is called the **Milky Way** and it is an edge-on view of a great disc of stars. The Milky Way goes all round our sky, so we are inside the disc. It is thinner in one direction, thicker in the other, so we are not even at the centre of the disc. The Milky Way is our view of our **galaxy**.

Milky Way

Andromeda Galaxy

Even further away are other galaxies made up of vast numbers of stars. The **Andromeda Galaxy** is over 2 million light years away. Our Milky Way Galaxy would look very similar from outside.

Millions of galaxies are seen further and further away in space. They make up the **Universe** which is perhaps 20 000 million light years across.

1 Try to match these names to the descriptions of planets **a** to **e** below:
Jupiter, the 'king god'; Mars, the god of war; Venus, goddess of love; Saturn, the old father of Jupiter, who walks very slowly; Mercury, the speedy messenger of the gods.
a This planet moves quite quickly, but is seen only just before sunrise or just after sunset.
b This planet is also near the Sun in the sky, but is the brightest planet.
c This planet is red, so it reminded people of blood.
d This planet is a golden orange, perhaps rather royal, they thought.
e This planet moves very slowly.
2 If light rays travel at 300 000 km/second, calculate how far 1 light year is in km.

6.4 The inner planets

The Sun is a star. It has nine major planets orbiting it. Many have smaller moons orbiting them. Like our Moon, other planets and moons are only seen because of the sunlight they reflect.

Moving out from the Sun, the four **inner planets** are Mercury, Venus, Earth, and Mars. They are smaller than most of the outer planets and are sometimes called the 'rocky dwarfs'.

Mercury

Although Mercury is quite bright, it is hard to see it in the sky because it is always close to the Sun. In many ways, Mercury is a larger version of our own Moon. It has a cratered surface, with some plains, and no atmosphere.

Diameter   4880 km   MERCURY
Mass 3.3×10^{23} kg
    (0.06 of Earth)
Typical surface
    temperature $-180\,°C$ to $380\,°C$
Atmosphere   None

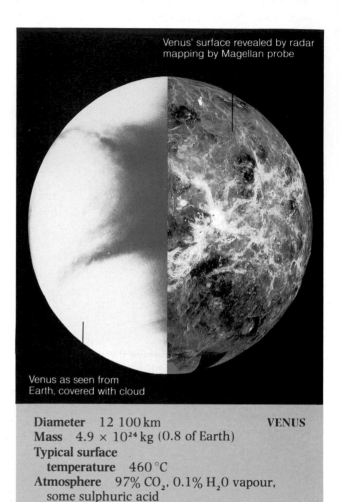

Venus' surface revealed by radar mapping by Magellan probe

Venus as seen from Earth, covered with cloud

Diameter   12 100 km   VENUS
Mass 4.9×10^{24} kg (0.8 of Earth)
Typical surface
    temperature $460\,°C$
Atmosphere   97% CO_2, 0.1% H_2O vapour,
    some sulphuric acid

Venus

Apart from the Sun and the Moon, Venus is the brightest object in the sky. During its orbit, it comes nearer Earth than any other planet, yet it has always been a place of mystery.

Venus is nearly as big as the Earth. Through a telescope it appears as a white disc or crescent, with no features. It is covered by clouds, made of water droplets. In the last fifty years scientists have found that Venus would be a very unpleasant place for humans.

Carbon dioxide makes up 97 per cent of the atmosphere. The remaining 3 per cent includes sulphuric acid. Venus suffers from an extreme greenhouse effect.

Spaceprobes have dropped into Venus' atmosphere to measure the temperatures and pressures. But the probes do not last long once they are on the surface. In 1990, the *Magellan* spaceprobe was put into orbit around the planet so that it could produce a detailed map of Venus' rocky surface. The map used radar, not ordinary photography.

Mars

Because of its surface colour, Mars is sometimes called the red planet. People once thought that it was inhabited. Through telescopes, some said they could see canals on its surface. However, when spaceprobes studied Mars, it was clear that the 'canals' were an optical illusion.

Mars has a thin atmosphere. The surface pressure is less than 1% of that on Earth. Even so, there are strong winds at times, and the dusty surface is blown about to make sandstorms.

Mars has features that may be evidence of natural processes similar to those on Earth.

Diameter	6780 km	**MARS**
Mass	6.5×10^{23} kg (0.11 of Earth)	
Typical surface temperature	$-120\,°C$ to $25\,°C$	
Atmosphere	95% CO_2, 3% N_2, traces of H_2O, Ar, O_2	

One Martian volcano is the largest known on any planet. Olympus Mons is 25 km high, 600 km in diameter, with a 65 km diameter crater. It is made of basalt lavas.

Mariner Valley is 2500 km long, up to 5 km deep, and from 100 to 200 km wide. Scientists are still wondering if Martian canyons were carved by water alone, or by giant flows of mud.

The problem of life on Mars was investigated by the Viking probes which landed in 1976. They sampled the soil and tested for signs of life. The results seemed to be negative.

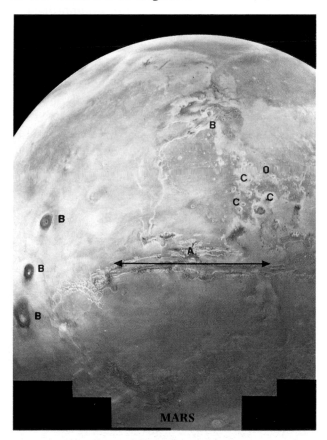

MARS

1 *The photographs in this Unit are not to scale.* List the four inner planets in order of size, with the largest first.
2 How does the landscape of Mercury appear to have been formed?
3 What properties do materials need if they are used to build a Venus probe?
4 Why is radar needed to map Venus?
5 Study the photograph above. It shows evidence of volcanoes, canyons, and craters on Mars. Match the letters to the features.

6.5 Gas giants and moons

In 1977, two *Voyager* spaceprobes were launched. Their mission was to pass close to Jupiter, Saturn, Uranus, and Neptune. These huge planets are called the **gas giants** as they are mainly gas and have no solid surface. Along with tiny Pluto, they make up the **outer planets**.

Jupiter

Voyager 1 reached Jupiter in 1979. As it passed **Io**, one of Jupiter's 16 moons, Linda Morabito at *Voyager* Control used a computer to improve one of its pictures. She saw a bright plume above Io. She had found the first active volcano beyond Earth.

Io's gravity is weak. Nearby Jupiter's enormous gravitational pull causes 'tidal bulges' 100 metres high on the moon's rocky surface. This generates heat and causes the volcanoes. Material from Io's volcanoes is ejected at around 1 km/s – about 20 times faster than from volcanoes on Earth.

Diameter 142 800 km	**JUPITER**
Mass 1.9×10^{27} kg	
(320 of Earth)	
Typical surface	
temperature $-120\,°C$	
Atmosphere 90% H_2, 4.5% He,	
traces of methane, ammonia	

Jupiter has huge weather systems high in its atmosphere. The **Great Red Spot** appears to be a storm that has raged for centuries. It is so big that the Earth would fit inside it.

In 1994, the fragments of a broken-up comet crashed into Jupiter's atmosphere, causing flashes of light, and leaving disturbances which astronomers were able to study and photograph.

Saturn

Saturn's rings are not one solid mass. They consist of billions of fragments of ice ranging in size from a few millimetres to several metres. Each fragment is a tiny 'moonlet' in its own orbit.

Saturn has 23 moons. The largest, **Titan**, is bigger than Mercury. It has a dense atmosphere, mainly of nitrogen, but mixed with methane and ethane. With the temperature at $-180\,°C$, these form pools which cause a hazy **photochemical smog**.

A volcano erupting on Io

Saturn and some of its moons

Diameter   120 000 km   **SATURN**
Mass 5.7×10^{26} kg
    (95 of Earth)
Typical surface
    temperature $-130\,°C$
Atmosphere   94% H_2, 6% He

Uranus

Using telescopes on Earth, astronomers had found five moons around Uranus, and nine thin rings. *Voyager 2* found ten more moons and at least two extra rings.

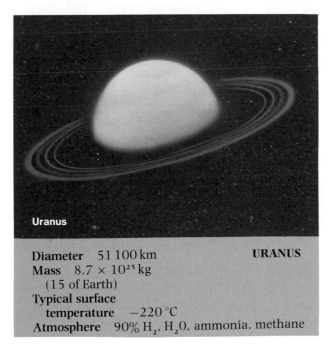

Uranus

Diameter   51 100 km   **URANUS**
Mass 8.7×10^{25} kg
    (15 of Earth)
Typical surface
    temperature $-220\,°C$
Atmosphere   90% H_2, H_2O, ammonia, methane

Uranus' atmosphere is mainly helium and hydrogen, but its blue-green colour is caused by the methane at higher levels. Winds at speeds of 360 km/hour (220 mph) are common on the planet.

Neptune

Voyager 2 sped past Neptune in 1989, around 12 years after its launch. Like Uranus, Neptune's colour is caused by methane. The planet also has faint rings and its own family of 8 moons.

Triton, the largest moon of Neptune, has a similar atmosphere to Titan (of Saturn). *Voyager 2* found evidence of 'volcanoes' erupting from Triton's surface, possibly driven by liquid nitrogen.

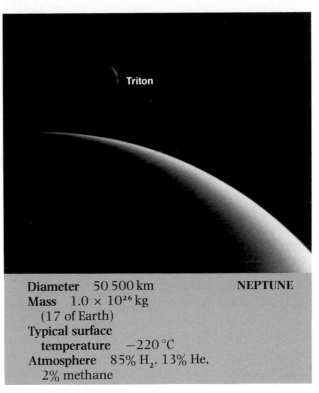

Triton

Diameter   50 500 km   **NEPTUNE**
Mass 1.0×10^{26} kg
    (17 of Earth)
Typical surface
    temperature $-220\,°C$
Atmosphere   85% H_2, 13% He,
    2% methane

1 *The photographs in this unit are not to scale.*
List the four gas giants in order of size, with the largest first.
2 Suggest reasons why volcanoes on Io eject material at a much higher speed than on Earth.
3 What do Saturn's rings consist of? What other planets have rings?
4 What gases might be found on Triton?
5 What problems would there be in sending a manned mission to the outer planets?

6.6 The Solar System

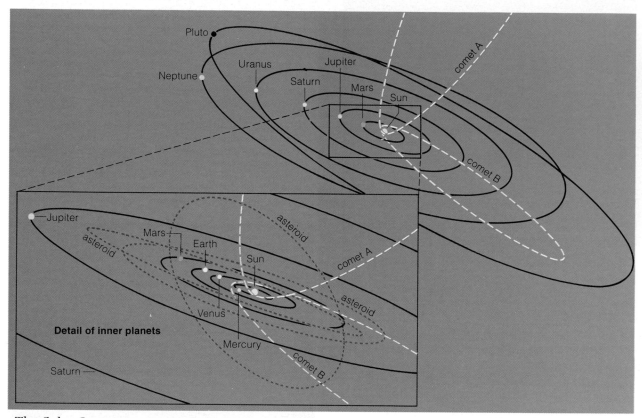

The Solar System

The Sun, its planets, and other objects in orbit are together known as the **Solar System**.

Between Mars and Jupiter are thousands of minor planets called the **asteroids**. The largest, **Ceres**, is only 1000 km across. Some asteroids stray further in or out. Mars has two small moons which may well be asteroids caught by Mars' gravity. Crashing asteroids are responsible for some of the impact craters on moons and planets.

Meteorites are probably fragments of asteroids that broke up. The iron meteorites came from the centres of these objects, and the stony meteorites from their outer parts.

Comets are lumps of 'dirty ice' a few kilometres across. They have highly elliptical orbits which can take them out beyond Neptune then back towards the Sun. Occasionally, they can crash into planets and moons and make craters.

Planet	Average distance from Sun in million km	Time for one orbit in years
Mercury	58	0.2
Venus	108	0.6
Earth	150	1.0
Mars	228	1.9
Jupiter	778	11.9
Saturn	1427	29.5
Uranus	2870	84.0
Neptune	4497	164.8
Pluto	5900	247.8

The gravitational pull on a comet is weakest when it is furthest from the Sun. The closer it gets, the faster it travels. When near the Sun, it warms up and gives off a long tail of dust and gas.

The origin of the Solar System

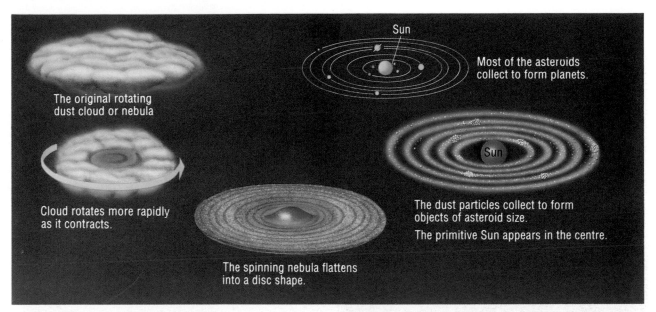

The original rotating dust cloud or nebula

Cloud rotates more rapidly as it contracts.

The spinning nebula flattens into a disc shape.

Most of the asteroids collect to form planets.

The dust particles collect to form objects of asteroid size.

The primitive Sun appears in the centre.

How the Sun and planets formed

Scientists think that the Solar System formed about 4500 million years ago in a huge cloud of gas and dust called a **nebula**. Gravity slowly pulled the material into blobs. One blob in the centre grew especially large and hot as more and more material crashed into it. Around it, smaller blobs orbited, growing bigger as they swept up most of the remaining material in the cloud. Worlds with no atmosphere, like our Moon, still show the large craters made in the later stages of this process.

The Sun

The Sun formed from the large blob at the centre of the cloud. In time, it became massive enough to collect most of the hydrogen gas around it and to pull smaller objects into orbit.

Deep inside the blob, the gases became more and more squashed. Eventually, their pressure and temperature became high enough to push hydrogen nuclei together to form helium nuclei. This process is called **nuclear fusion** and it releases a huge amount of energy. Once nuclear fusion had started, the blob had become a star – the Sun.

Scientists think that the Sun has enough hydrogen fuel left to keep it shining for another 6000 million years.

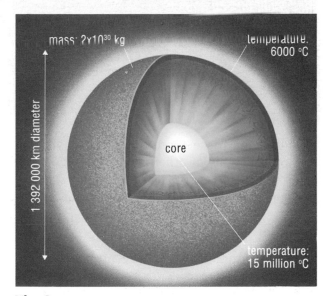

mass: 2×10^{30} kg

temperature: 6000 °C

1 392 000 km diameter

core

temperature: 15 million °C

The Sun

1 Which planet gets most heat (per m²) from the Sun? Why?
2 Look at the table on the left. Is there is a link between the time for one orbit and a planet's distance from the Sun? Explain your answer.
3 Which object in our Solar System has the strongest gravitational field?
4 By what process does the Sun get its energy?
5 What substance does the Sun use as its fuel?

6.7 The view from space

By launching rockets into space, it is possible to find out more about other planets and stars. But from space, it is also possible to observe and measure what is happening down on the Earth.

Satellites

There are hundreds of satellites in orbit around the Earth. Most are in circular orbits. (For more on orbits, see page 33.)

A satellite in a low orbit needs the highest speed. For example, a satellite at a height of 300 km, just above the Earth's atmosphere, must travel at 29 000 km/hour (18 000 mph) to maintain a circular orbit. At this speed, it will take 90 minutes to orbit the Earth.

A higher orbit needs a lower speed. For example, a satellite in an orbit at 36 000 km from the Earth's surface must travel at 11 000 km/hour. At this speed, it will take 24 hours to orbit the Earth.

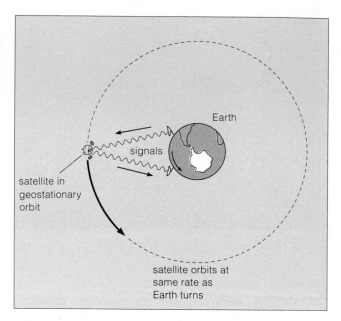

Communications satellites beam microwave signals from one part of the Earth to another. The signals can carry radio and TV programmes, computer data, and telephone conversations.

Satellites of this type are normally put into a **geostationary orbit** like the one shown above. This means that their orbital motion exactly matches the Earth's rotation so that they appear stationary in the sky. As a result, satellite TV dishes down on the ground can be mounted in fixed positions. They do not have to move in order to track the satellite. For a geostationary orbit, an orbit time of exactly 24 hours is required.

Navigation satellites are used by boats and planes to locate their position on the Earth's surface. The **Global Positioning System (GPS)** has a network of satellites which circle the Earth in geostationary orbits. They transmit synchronized time signals which can be picked up by a small receiver below. In the receiver, a tiny computer compares the arrival times of signals from different satellites. Knowing the speed of the signals and their time separation, it can calculate a position to within a few metres.

The planets and beyond?

Astronauts have travelled to the Moon using rockets like that in the diagram below. One day, they may reach Mars. But manned journeys to more distant parts of the Solar System are most unlikely in the foreseeable future.

One problem is gravity. Lifting even a small load into space requires a huge rocket and an enormous amount of fuel. The rocket engines have to be so powerful that they can only work for ten minutes or so before running out of fuel. For most of its journey, a spacecraft must 'coast'. And if it is to coast far enough into space not to be pulled back to Earth, it must reach a speed of at least 40 000 km/hour (25 000 mph) soon after launch. This is the **escape velocity** for the Earth.

Another problem is time. Distances in space are so vast, that a rocket leaving the Earth at 40 000 km/hour would take over ten years to reach Pluto. And a journey to the nearest star beyond the Solar System would take many thousands of years.

Monitoring satellites, like the weather satellite above, contain cameras or other detectors for scanning and surveying the Earth. Some are in low orbits which pass over the North and South Poles. As the Earth rotates beneath them, they can scan the whole of its surface.

Astronomical satellites, such as the Hubble space telescope, contain equipment for observing distant stars and galaxies. Unlike observatories on Earth, the signals they receive are not disrupted and weakened by the presence of the atmosphere.

1 Give four different uses of satellites.
2 If a satellite is to be put into a higher orbit, how will this affect the following:
a Its speed in orbit?   **b** The time for one orbit?
3 What is meant by a geostationary orbit? What are the advantages of putting a communications satellite into a geostationary orbit?
4 What problems would there be in sending a manned mission to Neptune?
5 Why will humans almost certainly never visit planets around other stars?

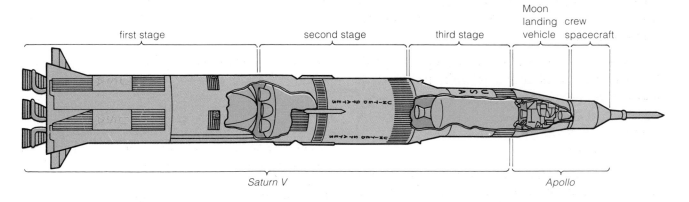

The *Saturn V* multistage rocket system, used to send the *Apollo* missions to the moon.

6.8 Beginnings and endings

Stars are formed in huge clouds of gas and dust called nebulae. The Sun was formed in a nebula about 4500 million years ago. But, like other stars, it will eventually run out of fuel and reach the end of its life cycle.

Death of the Sun

The Sun is powered by fusion, using hydrogen as its nuclear fuel. In about 6000 million years' time, it will have converted all its hydrogen to helium. Then it will swell to about 100 times its present diameter and its outer layer will cool to a red glow. It will have become a **red giant**.

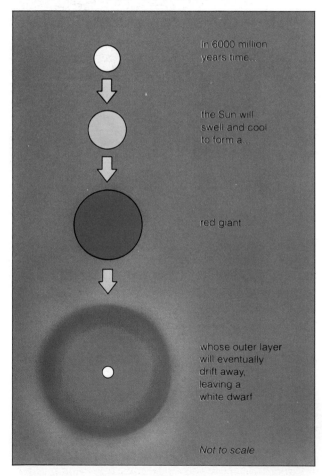

In 6000 million years time...

the Sun will swell and cool to form a...

red giant...

whose outer layer will eventually drift away, leaving a white dwarf

Not to scale

Eventually, the Sun's outer layer will drift into space, exposing a hot, dense core called a **white dwarf**. This tiny star will use helium as its nuclear fuel. When this runs out, the star will cool and fade for ever.

The Crab Nebula – the remains of a supernova

Supernovae and stardust

In each galaxy, new stars are forming and old ones are dying. But the most massive stars have a different fate from that of our Sun. Eventually, they blow up in a gigantic nuclear explosion called a **supernova**. This leaves a core in which matter is so compressed that electrons and protons react to form neutrons. The result is a **neutron star**.

If a neutron star has enough mass, it continues to collapse under its own gravity. Nothing can resist the pull. Even light cannot escape. The star becomes a **black hole**.

In stars, nuclear reactions change lighter elements into heavier ones. However, to make elements which are heavier than iron, the extreme conditions which create a supernova are needed. The Sun and inner planets contain very heavy elements. This suggests that the nebula in which the Solar System formed included remnants from an earlier supernova.

The expanding Universe

The light waves we receive from distant galaxies are 'stretched out' – their wavelengths are longer. This is called the **Doppler effect**. Scientists think that it occurs because the galaxies seem to be rushing apart at high speed. We are living in an expanding Universe.

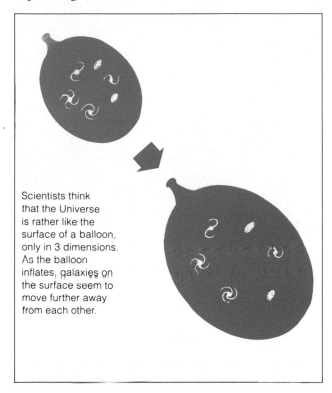

Scientists think that the Universe is rather like the surface of a balloon, only in 3 dimensions. As the balloon inflates, galaxies on the surface seem to move further away from each other.

The Universe may have been created in a gigantic explosion called the **big bang**. Here are two pieces of evidence to support this idea:

● Radio telescopes have picked up background microwave radiation from every direction in space. It may be an 'echo' of the big bang.
● As the galaxies seem to be moving apart, they may once have come from the same tiny volume of space.

Scientists have estimated the rate at which the galaxies seem to be moving apart. The value has not been agreed, but is thought to be about 1 km 1/15 000 km per year per million km of separation. They call this the **Hubble constant**. From it, it is possible to work out that the galaxies must have started to separate about 15 000 million years ago. If so, that is when the big bang occurred.

The gravitational pull between galaxies is gradually slowing the expansion of the Universe. But no one yet knows whether the expansion will ever stop. That depends on how much mass there is in the Universe.

Professor Stephen Hawking

Some of the most important work on the mathematics of black holes and the big bang has been carried out by Professor Stephen Hawking. He has done this in spite of almost total physical disability caused by a disease of the nervous system. Using state-of-the-art technology, including a voice synthesizer, he has prepared lectures and books which have changed our understanding of the Universe.

1 Explain what each of the following is:
nebula   red giant   white dwarf   supernova
2 How is a black hole formed?
3 Why is it likely that the Sun formed in a nebula containing material from an earlier supernova?
4 What evidence is there that the Universe may have started with a big bang?
5 What force is causing the expansion of the Universe to slow down?
6 The value of the Hubble constant is uncertain. If its value is 1/10 000 km per year per million km separation, how old is the Universe?

Questions on Section 6

1 The diagram below shows the Moon in orbit around the Earth, and the Earth in orbit around the Sun.

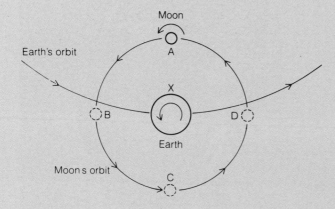

Not to scale

a

24 hours	28 days	365 days

Which of the above is closest to the time it takes:
   i The Earth to rotate once on its axis?
   ii The Earth to complete one orbit of the Sun?
   iii The Moon to complete one orbit of the Earth?
   iv The Moon to rotate once on its axis?

b One problem with diagrams of the Sun, Earth, and Moon, is that it is very difficult to show the various sizes and distances to scale on the same picture. The diagram above has not been drawn to scale correctly. Write down *three* things which you think are wrong with the diagram.

c On the diagram above, four positions of the Moon have been shown: A, B, C, and D. Say in which position (or positions) you think the following might occur, and give a reason for each answer.
   i A high tide at point X on the Earth.
   ii An eclipse of the Sun.
   iii An eclipse of the Moon.

2 The diagram below shows light from the Sun striking the Earth at one particular time of year.

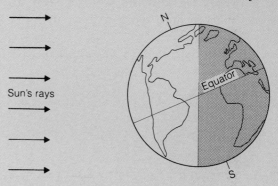

a

March	June	September	December

Which of the above months is the correct one for the diagram?

b Why, during the summer, is the daytime longer than the night-time?

c Why, when it is summer in the northern hemisphere, is it winter in the southern hemisphere?

3 The diagrams below show the constellation Taurus on two nights a month apart.

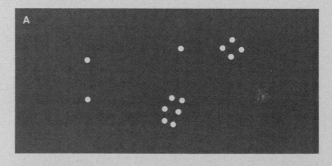

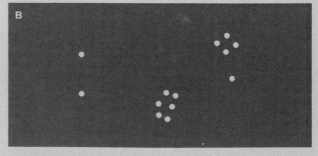

a What evidence is there that one of the objects is a planet? Make a sketch to show which object it is.

b Why will the planet not always appear to be among the stars of Taurus?

4 Here is some data about the planets:

Planet	Average distance from Sun in million km	Time for one orbit in years	Average surface temperature in °C
Mercury	58	0.2	350
Venus	108	0.6	480
Earth	150	1.0	22
Mars	228	1.9	−23
Jupiter	778	11.9	−150
Saturn	1427	29.5	−180
Uranus	2870	84.0	−210
Neptune	4497	164.8	−220
Pluto	5900	247.8	−230

a How would you expect a planet's surface temperature to depend on its distance from the Sun? Does the data support this?

b One planet suffers from an extreme greenhouse effect. Use the data to decide which planet this is. Explain your answer.

c Ceres is a minor planet (asteroid). It takes 4.6 years to complete one orbit of the Sun. Its orbit is approximately circular. By plotting a suitable graph, estimate the average distance of Ceres from the Sun.

5 The diagram below shows two identical satellites, A and B, in orbit around the Earth.

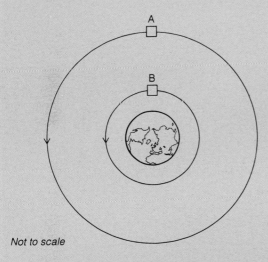

Not to scale

a Which satellite is pulled most strongly by the Earth's gravitational field?

b Which satellite has the highest speed?

c Which satellite will take longest to complete one orbit of the Earth?

d Satellite A is in a geostationary orbit. What does this mean?

e Why are communications satellites normally put into geostationary orbits?

6

meteorite	galaxy	supernova
big bang	constellation	black hole
moon	comet	Solar System

Which of the above is the best match for each of the following descriptions?

a A rocky object orbiting a planet.

b The Sun, its planets, and other objects in orbit.

c A small rocky object which collides with a planet, and may be a fragment from an asteroid.

d A clump of ice, gas, and dust, usually in a highly elliptical orbit around the Sun.

e A huge group of many millions of stars.

f A gigantic explosion that occurs when a very massive star has used up its nuclear fuel.

7 The Sun is a star. It formed in a nebula about 4500 million years ago. The diagram below shows what is likely to happen to the Sun in about 6000 million years' time.

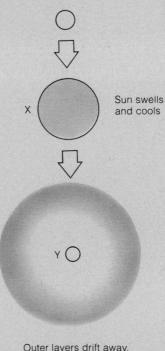

Sun swells and cools

Outer layers drift away, exposing a white-hot core

a What is a nebula?

b What makes matter in a nebula collect together to form a star?

c The Sun gets its energy from nuclear reactions that change hydrogen into helium. What is this process called?

d What type of star has the Sun become at X in the diagram above?

e What type of star is left when the core, Y, is exposed?

The scale of the Universe

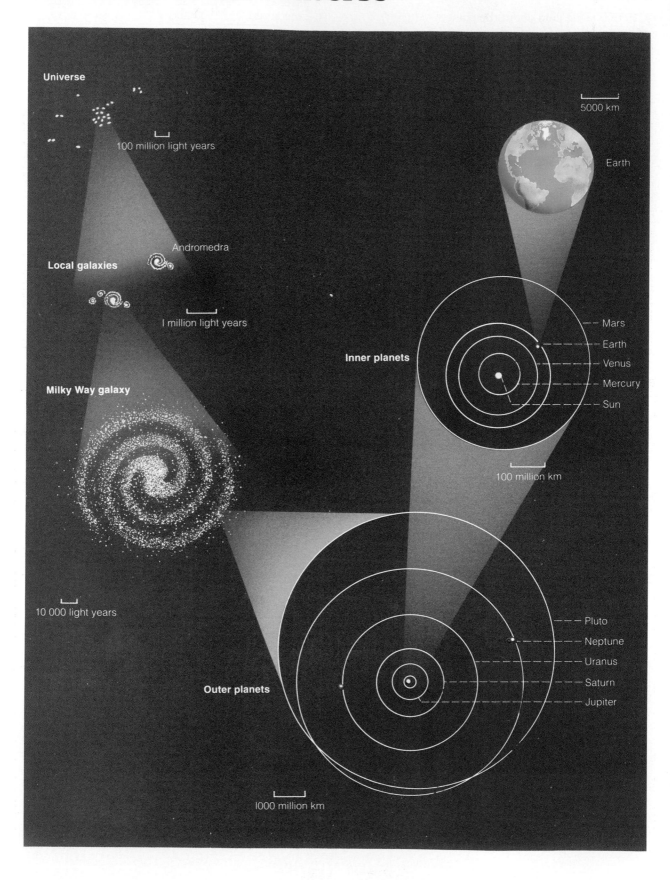

Units and symbols

Units and symbols		
Quantity	*Unit*	*Symbol*
mass	kilogram	kg
length	metre	m
time	second	s
force	newton	N
weight	newton	N
pressure	pascal	Pa
energy	joule	J
work	joule	J
power	watt	W
voltage	volt	V
current	ampere	A
resistance	ohm	Ω
charge	coulomb	C
temperature	kelvin	K
temperature	degree Celsius	°C

Bigger and smaller

To make units bigger or smaller, prefixes are put in front of them:

micro (μ) = 1 millionth = 0.000 001 = 10^{-6}
milli (m) = 1 thousandth = 0.001 = 10^{-3}
kilo (k) = 1 thousand = 1000 = 10^3
mega (M) = 1 million = 1 000 000 = 10^6

For example

1 micrometre = 1 μm = 0.000 001 m
1 millisecond = 1 ms = 0.001 s
1 kilometre = 1 km = 1000 m
1 megatonne = 1 Mt = 1 000 000 t

Electrical symbols

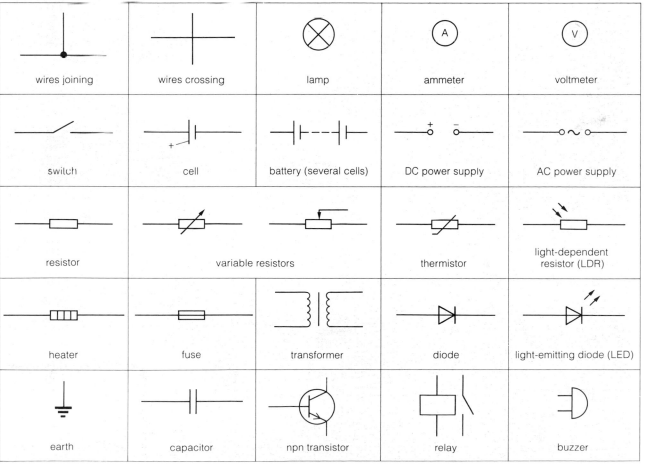

wires joining	wires crossing	lamp	ammeter	voltmeter
switch	cell	battery (several cells)	DC power supply	AC power supply
resistor	variable resistors		thermistor	light-dependent resistor (LDR)
heater	fuse	transformer	diode	light-emitting diode (LED)
earth	capacitor	npn transistor	relay	buzzer

Answers

1.1 (page 7)
3 1600 g, 1450 mm
4 **a** 1000 **b** 10 **c** 100 000
   d 100 **e** 1 000 000
5 **a** 1000 mm **b** 1500 mm
   c 1534 mm **d** 1.652 m
6 **a** 27 750 mm **b** 1600 m
   c 6500 mg **d** 1.5 m **e** 1.7 kg
7 kg, 100 cm
8 750 cm³, 0.75 l
9 24 cm³, 192 cm³

1.2 (page 9)
1 **a** 4000 kg **b** 2 m³ **c** 39 kg
2 **a** C **b** B **c** C
   A aluminium B concrete
   C steel
3 **a** 2 m³ **b** 4800 kg **c** 62 blocks

1.3 (page 11)
1 **a** 50 cm **b** 2.4 g/cm³
2 **a** 160 g **b** 0.8 g/cm³
3 A silver B gold C mixture

1.4 (page 13)
1 25 m/s
2 10 m, 50 m, 100 m, 9 s
5 30 m/s, 50 m/s
6 2.5 m/s²
7 4 m/s²

1.5 (page 15)
1 **c** 56 m, 8 m/s
   d 3 s, 36 m, 12 m/s
2 **a** 40 m/s **b** 20 s **c** 50 s
   d 40 m/s, 2 m/s, 2 m/s **e** 4 m/s²

1.6 (page 17)
1 **a** D **b** C **c** A
3 **a** 0.1 s **b** 5 **c** 0.1 s **d** 20 mm
   e 200 mm/s
   f 100 mm, 1000 mm/s
   g 800 mm/s **h** 800 mm/s²

1.7 (page 19)
1 **a** FALSE B TRUE C FALSE
3 10 m/s², 50 m/s, 10 m/s, 125 m,
   500 m
4 YES

1.9 (page 23)
2 **a** 4 N **b** 1 N **c** 8 N
3 5 N, 10 m/s²
4 **a** A and D **b** B **c** C
5 Honda, Boeing

1.10 (page 25)
2 20 N, 40 N, 5 N
5 **a** B **b** A and D **c** C **d** C
6 **a** 0.625 m/s² **b** 0.625 N/kg
   c 160 N

1.11 (page 27)
3 **c** 600 N **d** 60 kg

1.13 (page 31)
1 **a** 5000 kg m/s **b** M/cycle
   c Car
2 4 N
3 400 N
4 4 m/s

1.14 (page 33)
1 **a** 2 s **b** 50 m/s **c** 100 m
2 **a** i) MORE ii) MORE
   iii) MORE

1.16 (page 37)
1 **a** C **b** A and D **c** B
2 **a** 16 Nm **b** 12 Nm **d** 1 N

1.17 (page 39)
2 **b** 1 m **c** 100 Nm **d** 100 Nm
   e 100 N

1.18 (page 41)
2 **b** 40 mm **e** 3.9 N **f** 2.8 N

1.19 (page 43)
1 6 Pa
2 **a** 120 N, 140 N, 300 N
   b 1200 Pa, 700 Pa, 1000 Pa
3 No
4 **a** 30 N **b** 20 N

1.20 (page 45)
1 **a** 10 000 Pa **b** 20 000 N
2 **a** 100 000 Pa **b** 200 000 Pa
   c 300 000 Pa
3 7500 Pa, 0.75 m
4 500 000 Pa, 100 m, LESS

1.21 (page 47)
1 **a** 100 000 Pa **b** 200 000 Pa
   c 600 000 Pa, Yes, 70 m

1.22 (page 49)
1 **c** 760 mm Hg **d** 450 mm Hg
   e 1210 mm Hg
3 103 000 Pa

1.23 (page 51)
1 **b** 1000 N
   c 1000 N
3 **a** Yes **b** No

Questions on Section 1 (pages 52–53)
1 **a** Jane **b** Emma
   c 5 s **d** 7 m/s
   f 1.5 m/s² **g** 12 m
2 **a** 10 **c** 3 s
   d iii) Downward iv) Equal
3 **a b c e** High **d f** Low
4 **c** 200 Nm **d** 200 Nm
   e 0.5 m **g** 200 N
5 **a** 12 mm
   c i) 6 mm ii) 9 mm
6 **b** 16 000 N **c** 16 000 N
   d 2000 Pa
7 **a** 12 N **b** 15 m/s
   c i) 60 kg/ms ii) 60 kg m/s
   iii) 7.5 m/s

2.1 (page 55)
1 **a** 18 J **b** 6 J **c** 0.1 J

2.2 (page 57)
2 3500 J

2.3 (page 59)
1 **a** 32 J, 1 J **b** 5 J, 10 J
   c 37 J, 11 J
2 **a** 100 000 kg **b** 8000 m/s
   c 100 000 m **d** 100 000 MJ
   e 3 200 000 MJ **f** 3 300 000 MJ

2.4 (page 61)
1 **a** C **b** 2 h, 2 h, 4 h
   c A £36, B £38, C £52
   d A £68, B £66, C £96
   e C **f** A, B

2.5 (page 63)
1 **a** 1000 J **b** 10 s **c** 100 J
2 **a** 500 N **b** 10 000 N **c** 250 W

1.14 (page 33)
1 **a** 2 s **b** 50 m/s **c** 100 m
2 **b** i) MORE ii) MORE
   iii) MORE

2.7 (page 67)
2 **a** 100 N **b** 100 000 Pa
   c 100 000 Pa **d** 200 N
   e MORE **f** MORE

2.9 (page 71)
2 **a** B **c** E
   d A **f** A
3 **c** X 2000 MW; Y 1500 MW
   e X 36%; Y 27%

2.11 (page 75)
4 0.000 0015 mm

2.12 (page 77)
1 **a** 100 °C **b** 373 K **c** −273 °C
   d 0 K **e** 0 °C **f** 273 K
3 **b** 313 K **c** 4 h **d** 10:30 a.m.

2.14 (page 81)
3 0.1 m
4 YES

2.18 (page 89)
2 **a** 8000 mm Hg **c** 1300 mm Hg
3 **a** 300 K
   d 8 atm

2.20 (page 93)
3 **b** 12 °C **c** 7 min

2.21 (page 95)
2 **a** 105 000 J

2.23 (page 99)
2 330 000 J, 660 000 J, 3 300 000 J
3 **b** 3 min C 6 min **d** 53 °C

2.24 (page 101)
3 **a** 2 300 000 J **b** 2300 J
   c 1000 s
4 Alps 95 °C, sea 103 °C, tap 100 °C

Questions on Section 2 (pages 106–107)
1 **a** C **b** B **c** D **d** A
2 **d** 150 J **e** 150 000 J
   f 75 000 J **g** 75 kW
4 **c** 1000 Pa *d* 1000 Pa
   e 200 N
5 **a** Evap. **b & c** Conv.
   d Cond. **e** Conv.
6 **c** i) 1000 ii) 3000 J
   iii) 1 260 000 J
   d i) 4200 J ii) 420 000 J
   iii) 3 °C
7 **a** In kPa: 100, 200, 300, 400
   b 20 m **c** 2 m³

3.1 (page 109)
2 300 000 km/s

3.6 (page 119)
2 **b** 5 **c** 1 **d** 60 cm

3.7 (page 121)
1 D, A

3.8 (page 123)
1 **b** 20 mm, 6 mm 15 mm
2 Wave 1: 32 m/s
   Wave 2: 32 m/s, 2 m
   Wave 3: 32 m/s, 32 Hz
3 **a** B **b** A **c** C

3.10 (page 127)
3 3 m

252

3.11 (page 129)
3 a B b C

3.12 (page 131)
1 Red, green, blue
2 White 3 Yellow
4 Red, green, blue
5 All 6 Blue; red & green
7 Green; red & blue 8 Black

3.14 (page 135)
2 a 10 mm

3.16 (page 139)
1 Aircraft, meteorite
2 a 330 m b 660 m c 3300 m
 d 33 m
4 a 0.1 s b 140 m

3.17 (page 141)
3 a Trumpet b Keyboard, flute
 c 600 Hz

3.19 (page 145)
1 P 2 Surface waves
3 a P b S
4 a Love
5 10 km

3.20 (page 147)
2 b Record b at Y
3 a B

Questions on Section 3 (pages 148-149)
2 a Infrared b X-rays
 c i) Gamma ii) Radio
 iii) Radio
 iv) Ultraviolet
 d i) Red ii) Violet
3 e 2 kW f 5 m²
4 a A b C c A
5 a ii) Convex
6 c D d C
 e 1.5 m f 440 Hz
7 a i) 20 kHz ii) Y

4.2 (page 153)
3 a 60 000 J b 3 h c £1

4.3 (page 155)
1 X 1.8 A, Y 0.76 A
3 3 A, 3 A, 3 A
4 a 40 C b 40 s c 5 A

4.4 (page 157)
1 a C b A
2 4 V, 2 V
3 a 6 V b 9 J c 3 J
4 a B b A c 1000 C
 d 12 000 J

4.5 (page 159)
1 23 Ω
3 a 4.4 A b 1.2 Ω c 1.9 Ω
 d 2.4 Ω

4.7 (page 163)
1 B SERIES, all OFF
 C PARALLEL, others ON
2 A and G, B and H, C and E,
 D and F

4.8 (page 165)
1 a 6 Ω b 15 V c 3 A d 0.5 A
 e 2 V
2 a 3 A b 6 A c 9 A d 2 Ω
 e 2 Ω
3 a 2 A b 6 A

4.10 (page 169)
1 a 1150 W b 11
2 A 460 W, B 690 W, C 92 W
3 a A 0.46 kW, B 0.92 kW

C 1.15 kW, D 0.023 kW,
E 0.046 kW
 b A 2 A, B 4 A, C 5 A,
 D 0.1 A, E 0.2 A
 c A 3 A, B 13 A, C 13 A
 D 3 A, E 3 A
4 a 2 A b 12 W c 4 A d 48 W

4.12 (page 173)
1 J, kJ, kW h
2 b 3 600 000 J
3 A 0.5 kW h, B 1.5 kW h, C 3 kW h,
 D 24 kW h
4 Donna's father
5 a 100 p b 10 p c 2 p
6 21 p
 a 756 000 J b 6 min c 2.1 p

4.15 (page 179)
2 a A b C c B

4.17 (page 183)
3 c B-to-A d TOP

4.19 (page 187)
1 B-to-A

4.21 (page 191)
1 a C b D c B d B e A
2 a 10:1 b 46 W c 46 W
 d 0.2 A

4.22 (page 192)
2 b 5 c Extown, Oldwich
 d 50 MW e 150 MW
 f 100 000 000 J g 5000 A
 h 250 A

Questions on Section 4 (pages 196-197)
1 a i) YES ii) NO iii) NO
 iv) YES v) YES
3 a C b B
4 a ii) 4 A iii) 2 A b ii) 3 V
5 a 3 A b 12 W
 c 6 J d 2 C
6 a i) S₁ ii) S₁ and S₂
 b 8.5 A c 13 A
 f Blue g Brown
7 a 0.01 kW, 0.2 kW
 b 20 kW h d £8
8 a Electromagnet
 c Transformer

5.1 (page 199)
2 Sodium-23: 11 p, 12 n, 23
 Aluminium-27: 13 p, 14 n, 27
 Strontium-90: 38 p, 52 n, 90
 Cobalt-60: 27 p, 33 n, 60
3 a $^{226}_{88}Ra$ b $^{235}_{92}U$ c $^{16}_{8}O$ d $^{12}_{6}C$
4 A lithium-7, B thorium,
 3 lithium-6, D boron, A and C

5.4 (page 205)
3 A gamma, B alpha + gamma,
 C beta

5.5 (page 207)
1 Radium-226
2 c 4 min
3 b 50 s c 50 s d 50 s
4 a 8 days

5.7 (page 211)
2 15 mm, 20 mm a 30 V
 b 100 ms

5.8 (page 213)
3 a B b A c D; 8 V

5.9 (page 215)
1 a A b A c B

5.10 (page 217)
1 a D b A c C d B

5.11 (page 219)
1 a E, C, A, D

5.12 (page 221)
1 a A b C c A d D

5.13 (page 223)
1 a A B C Q
 0 0 1 0
 0 1 1 1
 1 0 0 0
 1 1 0 0
2 a NAND OFF, NOR OFF
 b NAND ON, NOR ON
 c NAND ON, NOR OFF

5.14 (page 225)
4 A and C

5.15 (page 227)
 a 2, 3, 6, 7, 6, 2, 2, 3
 b 010, 011, 110, 111, 110, 010,
010, 011

Questions on Section 5 (pages 230-231)
1 a - b Proton c +
 d 4 e Electrons
2 b GM tube
3 a Fission
 f i) 400 ii) 200 iii) 50
5 a LED b Relay c Thermistor
 d LDR e Diode
7 a Q: 0, 1, 1, 1 b OR
8 a B b Analogue c Digital

6.1 (page 233)
1 a Year b Day
2 A
4 a Dec 21 b June 21
5 12 h 6 Dec 21 7 June 21
8 B, D
9 A FALSE B TRUE C TRUE

6.2 (page 235)
1 27 days 4 2 weeks

6.3 (page 237)
1 a Mercury b Venus c Mars
 d Jupiter e Saturn
2 9.5 × 10¹² km

6.4 (page 239)
1 Earth, Venus, Mars, Mercury

6.5 (page 241)
1 Jupiter, Saturn, Uranus, Neptune

6.6 (page 243)
1 Mercury 3 Sun
4 Fusion 5 Hydrogen

6.7 (page 245)
2 a Less b More

6.8 (page 247)
6 10 000 million years

Questions on Section 6 (pages 248-249)
1 a i) 24 h ii) 365 d
 iii) 28 d iv) 28 d
 c i) A&C ii) A iii) C
2 a June
4 b Venus c 400 million km
5 a B b B c A
6 a moon b Solar System
 c meteorite d comet
 e galaxy f supernova
7 c Fusion
 d Red giant e White dwarf

Index

If more than one page number is given, you should book up the **bold** one first.

Acknowledgements

The publisher would like to thank the following for their kind permission to reproduce the following photographs:

p8 Royal Observatory, Edinburgh, **p10** Barnaby's Picture Library (right), British Geological Survey (middle right), Civil Aviation Authority (left), Ind Coope Burton Brewery Ltd (middle left), **p15** Zefa Photographic Library, **p16** The Churchill Group/Penny Giles, **p20** Alton Towers Ltd (middle left, bottom left, middle right), Colorsport (top right), Crown Copyright (bottom right), **p21** Volvo Concessionaires Ltd (centre), Derrick D Bryant (bottom right), **p22** Rolls Royce plc, **p23** British Petroleum (top left), J Allan Cash (bottom right), Porsche Cars Great Britain Ltd (right middle), **p26** Allsport (UK) Ltd (bottom left), Ford (UK) Ltd (bottom right), NASA (top right), **p27** Allsport (UK) Ltd, **p30** The Image Bank, **p33** Sporting Pictures (UK) Ltd, **p34** Allsport (UK) Ltd (top left), Colorsport (top right), **p35** Allsport (UK) Ltd (top), Zefa Photographic Library (bottom), **p38** Allsport (UK) Ltd, **p40** Environmental Picture Library (top right), Photographer's Library (top left), **p44** Robert Harding Picture Library (top), Ministry of Defence (centre), **p47** Shell/Esso, **p48** TV AM Enterprises (top left), Zefa Photographic Library (top right), **p50** Natural History Photographic Agency, **p51** The Image Bank, **p56** Colorsport, **p58** NASA, **p60** National Motor Museum, **p63** Ardea London Ltd, **p64** Colorsport, **p66** Hitachi (bottom right), **p69** Zefa Photographic Library, **p73** Central Electricity Generating Board, **p74** Zefa Photographic Library, **p76** Robert Harding Picture Library, **p79** British Steel (top left), **p80** British Airways, **p84** Schweppes (top left), **p85** Robert Harding Picture Library (bottom), Zefa Photographic Library (top), **p88** J Allan Cash Ltd, **p90** Barnaby's Picture Library (top left), Solarfilma (bottom right), **p92** Allsport (UK) Ltd (top left), Derek Fordham (bottom right), Frank Lane Picture Agency Ltd (centre right), **p93** J Allan Cash Ltd, **p94** J Allan Cash Ltd, **p96** Bubbles (bottom right), Sally & Richard Greenhill (bottom left), **p97** British Petroleum, **p98** B & C Alexander (right), Coca-Cola Great Britain (left), **p102** Allsport (UK) Ltd (left), Frank Lane Picture Agency Ltd (right), **p103** J Allan Cash Ltd (centre), Robert Harding Picture Library (centre right), **p104** Centre for Alternative Technology (top right), Susan Griggs/Comstock (top left), Robert Harding Library (bottom right), **p105** Oxfam/J Hartley, **p108** Casio Electronics Ltd (left), Robert Harding Picture Library (bottom right), Oxford Lasers, Oxford UK (centre right) **p109** Barclays Bank Plc (top left), Oxford Lasers, Oxford UK (bottom left), Spectrum (bottom right) **p113** Ardea London Ltd (centre left), Civil Aviation Authority (top right), Rankin Glass Ltd (centre right), **p116** British Museum/National History Museum, **p117** Robert Harding Picture Library, **p118** Argos (left), **p120** Argos (centre), **p122** Frank Lane Picture Agency Ltd, **p128** Austin Rover Group Ltd (bottom left), J Allan Cash (top left), Colorsport (top right), **p129** Banbury Homes and Gardens/Insight Marketing (centre), **p130** Science Photo Library, **p132** Sporting Pictures/G Waugh, **p133** Casio Electronics Ltd, **p135** SPL (left), **p136** Environmental Picture Library (top left), TRH Photo Library (top right), Zefa Photographic Library (bottom left), **p137** Argos (bottom middle), Bruce Coleman Ltd (bottom right), Philips Consumer Electronics (bottom left and top right), **p138** Frank Lane Agency Ltd, **p139** Odean Cinemas, **p140** CBS Press (centre), WEA (left), **p141** Roland (UK) Ltd, **p142** Civil Aviation Authority (top right), Picturepoint (top left), **p143** SPL, **p144** SPL, **p152** Zefa Photographic Library (middle), Fiona Corbridge (top left), **p156** Ardea London Ltd, **p160** Central Electricity Generating Board (top right), Frank Lane Picture Agency Ltd (bottom left), Network South East (centre), **p162** Fiona Corbridge, **p172** Braun Electric (UK) Ltd, **p175** Boxmag Rapid (right), **p180** Boxmag Rapid (top right), J P Browett (bottom right), **p186** Goodmans Industries Ltd (bottom middle), Hohner/Michael Kommer Associates (bottom right), **p188** Lucas (centre), Unibike (top left), Central Electricity Generating Board (bottom left), **p192** CEGB, **p194** J Allan Cash (bottom left), CEGB (top right), Milk Marketing Board (centre left), Zefa Photographic Library (bottom right), **p195** North Scotland Hydroelectric Board (centre right), North London Waste Authority (top left and right), **p198** Susan Griggs/Comstock (bottom), **p200** British Nuclear Fuels (top left), United Kingdom Atomic Energy Authority (top right), **p202** Tass (top left), UKAEA (top right), **p203** Gilbert Films (top right), **p204** UKAEA (top left), **p205** Hammersmith Hospital (bottom right), UKAEA (top right and left), **p206** The Manchester Museum, **p207** British Nuclear Fuels, **p208** UKAEA, **p209** British Nuclear Fuels, **p218** Argos (right), **p219** Hotpoint, **p224** Morphy Richards, **p226** SPL (bottom), **p228** Crafts Council (top right), Ford (UK) Ltd (top left and bottom left), Walk Fund (bottom right), **p229** MARS (bottom left), Massey Ferguson Holdings Ltd (centre right), Adrian Meredith (top), **p234** SPL, **p237** SPL, **p238** SPL, **p239** SPL (right), **p240** TRH Pictures, **p241** NASA (right), Photo Library International Ltd (bottom left), TRH Pictures (top left), **p244** SPL, **p245** ESA/Meteosat Data Service, **p246** TRH Pictures, **p247** Manni Mason's Pictures.

Additional photographs are by Peter Gould and Chris Honeywell.
With special thanks to F C Bennett & Sons Ltd, Smiths Security Services, The Straw Hat Bakery and N J Thake Cycles.

The illustrations are by:

Jeff Edwards, Clive Goodyer, Nick Hawken & Associates, and Jan Lewis.

Cover illustration by Danny Jenkins.